WATER QUALITY AND STANDARDS

Editors: Shoji Kubota and Yoshiteru Tsuchiya

Eolss Publishers Co. Ltd.,
Oxford, United Kingdom

Copyright © 2009 EOLSS Publishers/ UNESCO

Rev 1.1, February 2010

Information on this title: www.eolss.net/eBooks

ISBN - 978-1-84826-031-3 (e-Book Adobe Reader)
ISBN - 978-1-84826-481-6 (Print (Color Edition))

British Library Cataloguing-in-Publication Data
A catalogue record of this publication is available from the British Library.

Library of Congress Cataloging-in-Publication Data
A catalog record of this publication is available from the library of Congress

Singapore

WATER QUALITY AND STANDARDS

This volume is part of the set:

Water Quality and Standards Volume I
ISBN- 978-1-84826-030-6 (e-Book Adobe Reader)
ISBN- 978-1-84826-480-9 (Print (Full Color Edition))

Water Quality and Standards Volume II
ISBN- 978-1-84826-031-3 (e-Book Adobe Reader)
ISBN- 978-1-84826-481-6 (Print (Full Color Edition))

The above set is part of the Component Encyclopedia of *WATER SCIENCES, ENGINEERING AND TECHNOLOGY RESOURCES,* in the global *Encyclopedia of Life Support Systems (EOLSS),* which is an integrated compendium of the following Component Encyclopedias:

1. *EARTH AND ATMOSPHERIC SCIENCES*
2. *MATHEMATICAL SCIENCES*
3. *BIOLOGICAL, PHYSIOLOGICAL AND HEALTH SCIENCES*
4. *SOCIAL SCIENCES AND HUMANITIES*
5. *PHYSICAL SCIENCES, ENGINEERING AND TECHNOLOGY RESOURCES*
6. *CHEMICAL SCIENCES ENGINEERING AND TECHNOLOGY RESOURCES*
7. ***WATER SCIENCES, ENGINEERING AND TECHNOLOGY RESOURCES***
8. *ENERGY SCIENCES, ENGINEERING AND TECHNOLOGY RESOURCES*
9. *ENVIRONMENTAL AND ECOLOGICAL SCIENCES, ENGINEERING AND TECHNOLOGY RESOURCES*
10. *FOOD AND AGRICULTURAL SCIENCES, ENGINEERING AND TECHNOLOGY RESOURCES*
11. *HUMAN RESOURCES POLICY, DEVELOPMENT AND MANAGEMENT*
12. *NATURAL RESOURCES POLICY AND MANAGEMENT*
13. *DEVELOPMENT AND ECONOMIC SCIENCES*
14. *INSTITUTIONAL AND INFRASTRUCTURAL RESOURCES*
15. *TECHNOLOGY, INFORMATION AND SYSTEM MANAGEMENT RESOURCES*
16. *AREA STUDIES (REGIONAL SUSTAINABLE DEVELOPMENT REVIEWS)*
17. *BIOTECHNOLOGY*
18. *CONTROL SYSTEMS, ROBOTICS AND AUTOMATION*
19. *LAND USE, LAND COVER AND SOIL SCIENCES*
20. *TROPICAL BIOLOGY AND CONSERVATION MANAGEMENT*

CONTENTS

Preface   xv

VOLUME II

Inorganic Chemicals Including Radioactive Materials in Waterbodies   **172**

Yoshiteru Tsuchiya, *Department of Applied Chemistry, Faculty of Engineering, Kogakuin University, Japan*

Microbial/Biological Contamination Of Water   **194**

Yuhei Inamori, *National Institute for Environmental Studies, Tsukuba, Japan*
Naoshi Fujimoto, *Faculty of Applied Bioscience, Tokyo University of Agriculture, Tokyo, Japan*

VOLUME I

Water Quality Standards, and Monitoring   **74**
Yashumoto Magara, *Professor of Engineering, Hokkaido University, Sapporo, Japan*

Basic Concepts and Definitions in Water Quality and Standards   **105**
Yashumoto Magara, *Professor of Engineering, Hokkaido University, Sapporo, Japan*

Classification of Water Quality Standards   **122**
Yashumoto Magara, *Professor of Engineering, Hokkaido University, Sapporo, Japan*

Assessment of Standards   **135**
Masanori Ando, *Musashino University, Japan*

Groundwater Monitoring   **177**
Masanori Ando, *Musashino University, Japan*

Management of Water Supplies after a Disaster   **292**
Yashumoto Magara, *Professor of Engineering, Hokkaido University, Sapporo, Japan*
Hiroshi Yano, *Director, Kobe Municipal Water Works Bureau, Kobe, Japan*

Preface

Water is used ubiquitously; it is essential to maintain life on Earth, such as humans and animals, plants, and microorganisms.

Moreover, water is indispensable in agriculture use, fisheries, and industrial production for humans to lead a comfortable life. Furthermore, in the global environment, it is necessary to maintain the ecosystem in coexistence. Water on Earth can be classified into water contained in seawater, rivers, lakes and marshes, underground water, soil water, atmospheric water, and living bodies, and 97% is seawater. This water contains various elements in each source and shows characteristic water qualities. Therefore, water quality standards or guidelines are needed to maintain the water quality corresponding to the purpose, such as for drinking, agriculture, fisheries and industry. In addition, when water is polluted by human discharge into water environment, it should be improved, following standards to maintain the ecosystem in the aquatic environment.

This paper introduces water quality and the water quality standards required for the purpose of use. This paper consists of five topics: the necessity to conserve water quality, water quality standards, and monitoring are shown in the first section, and pollution sources and contaminants which worsen water quality by human activities are shown the in second section. Moreover, the cause and measurement of pollution load on the environment by microorganisms or chemical substances are described. In *"Water quality standards and monitoring"*, the evaluation method (for instance, TDI is used for the health effects on humans, and PEC/PNEC is used for the effects on the ecosystem) to set the water quality standard, and the basic concept of classification, definition, the state of the water quality of natural water, and water quality monitors of surface water and groundwater are described. In *"Water quality needs and standards for different sectors and uses"*, the preferable water quality for human health, agriculture, fisheries, the ecosystem, industrial use, and disaster emergencies are described. In *"Effects of human activities on water quality"*, water resources and the water cycle on the Earth, water use and their balance, reduction of the environmental load by the self-cleansing power of the water environment, and the cause and substance decrease of the water quality of groundwater and surface water after human activity are described. In *"Pollution sources"*, various pollution sources that influence water quality are classified by the influence of salinity, point sources, non-point sources, and soil, water pollution in cities with drainage, and agriculture, classified into water pollution by waste water treatment plant effluent, are shown. Finally, in *"Contamination of Water Resources"*, the source of contaminants such as organic substances, inorganics, radioactive materials, microbes and aquatic organisms are described. Also, physicochemical pollution such as turbidity, color, unpleasant taste and odor, and pH are described.

Using these topics, appropriate water quality and appropriate water quality standards according to the purpose of use, and the cause of reduced water quality, the situation, and measures were clarified.

The contribution of all those who participated in the preparation and finalization of *Water Quality and Standards* is gratefully acknowledged.

Yoshiteru Tsuchiya and Shoji Kubota
Editors

EFFECTS OF HUMAN ACTIVITIES ON WATER QUALITY

Koichi Fujie
Department of Ecological Engineering, Toyohashi University of Technology, Japan

Hong-Ying Hu
ESPC, Department of Environmental Science and Engineering, Tsinghua University, China

Keywords: Water pollution, modern history, human activity, chemical management, PRTR.

Contents

1. Modern history of water pollution
2. Countermeasures for water pollution control
3. Present situation of water quality
Glossary
Bibliography
Biographical Sketches

Summary

The effect of human activities on natural water quality is discussed by looking back at the history of water pollution in Japan. The countermeasures for water pollution control such as enactment of ordinances and laws, establishment of chemical management and control system are briefly introduced. In addition, a new concept of emission minimization system is proposed.

1. Modern history of water pollution

It is well known that natural aquatic systems have a capacity to detoxify a certain quantity of pollutants discharged into them. This phenomenon is called self-purification. A water body will be polluted when the pollutants discharged into it excede their capacity of self-purification. The source of pollutants in a natural water body can be divided into two groups: those from natural sources (such as soils, forests, etc.) and artificial sources, namely human activity. By looking back on the history of water pollution, it is easy to understand that the expansion of human activity is the main reason for water pollution.

Water pollution problems occurred even before the industrial revolution. One of the early records of river water pollution in Japan was the outflow of pit water from Ashio Copper Mines into the Watarase River during the late nineteenth century. The outflow of pit water damaged the river water and paddyfields on the riverside. With the growth of industrial activities, the volume of wastewater flowing into rivers increased, bringing serious water pollution problems in various areas of Japan. During the period of industrial reconstruction after World War II, fisheries had been greatly damaged by the discharge of industrial wastewaters. In the 1960s, during the period of rapid economic growth, water pollution became more widespread and serious in Japan.

For example, until around 1945 the Sumida River had clear water with good populations of fishes, and it provided a place for recreation for local people and also a source of income for local fishermen. However, fish could not survive in the river after about 1955 and offensive odors were produced by the dirty river. A survey of Shizuoka Prefecture (Japan) in 1961 reported that an area of 14 km^2 in and around the Tagonoura Bay was polluted by black sludge, indicating a serious pollution situation. In those days, water pollution was aggravated mainly by the wastewater discharged from factories and business establishments. In 1958, a paper mill discharged its semi-chemical pulp wastewater to the Edogawa River without any treatment, and this caused extensive damage to the shellfish cultures downstream. Many angry fishermen rushed to this paper mill and accused the manager of the plant of discharging wastewater to the river.

The pollution of the rivers flowing through large cities and industrial areas such as the Yodogawa River was further aggravated after about 1960. In 1962, an electroplating plant located at the area of the Tama River discharged huge amount of cyanide compounds to the river. The bodies of many fish such as ayu (sweet fish) were subsequently found to be floating in the river.

The water pollution also began to affect human health. The Minamata disease incident is one of the most famous examples in the world of damage to people's health as a result of water pollution. The Chisso Corporation had established a factory in Minamata City in 1908. From that time the wastewater from this factory was discharged to Minamata Bay and the water of the bay area was seriously polluted by the wastewater. By 1955, bodies of fish were found to be floating on the water surface of the bay and even cats and pigs in the bay area began to die of an unknown and strange disease. In 1956, a patient with brain damage of unknown origin was reported to the local public health center, and the so-called Minamata disease was officially identified. Quite a long time after the first report of Minamata disease, it was determined that the disease was caused by the ingestion of fish and shellfish polluted by wastewater from the factory. Organo-mercury accumulated in the fish and shellfish was taken into human body and this damage the nervous system.

Itai-itai disease is anther well-known example of the affect of water pollution on people's health. The water of the Zinzu River (Japan), which was polluted by metals such as cadmium, lead and zinc, had been used to irrigate paddy fields since the Taisho era (1912 to 1925) and this caused serious damage to local agriculture and human health. Patients appeared with a strange disease in this area. They were suffering terrible pains and they frequently cried "itai-itai" ("ouch-ouch"; *itai* means "sore" in Japanese), so the disease was named "itai-itai disease". The Itai-itai disease was reported to the medical academy in 1955, and after intensive research it was officially announced in 1968 that the origin of the disease was cadmium contained in the wastewater from a mineral mining company located upstream on the Zinzu River.

2. Countermeasures for water pollution control

Ordinances and laws for water pollution control. Ordinances and laws are indispensable for water pollution control. In the 1950s, in response to the many concerns raised about industrial pollution, many local governments of Japan enacted ordinances to have companies take measures for water pollution control. In 1958, the central government of Japan enacted two laws on water pollution control: "Water Quality Conservation Law" and "Factory Effluent Control Law". A Water Quality Standard was established for the Edogawa River based on the above two laws. Water Quality Standards for the Yodo, Kiso, Ishikari, and Ara rivers were also established soon after the enactment of these two laws. In 1967, the Japanese government enacted the Basic Law for Environmental Pollution Control, to promote comprehensive measures against various forms of environmental pollution. In Japan, environmental administrations have been conducted mostly under the framework of the Basic Law for Environmental Pollution Control, and they have been very successful in water pollution control. Most of the approaches for water pollution control under the framework of the basic law at that time were based on the command-and-control mechanism. This proved very costly, however, and not sufficient to deal with new problems of environmental pollution. Based upon this situation, the Japanese government enacted the Environment Basic Law in 1993; this provides the philosophy for environmental policies, the direction for basic actions and a general framework for environmental policies. In 1994, under the Environment Basic Law, the Japanese Cabinet established the Basic Environment Plan to provide fundamental directions for environmental protection. The Basic Environmental Plan stresses that for the protection of the aquatic environment the incoming pollution load must be kept within the purification capacity of the natural water bodies. When planning action for protection of the water environment, all aspects of water quality and quantity, aquatic organisms and watersides should be taken into consideration. In addition to human health and aquatic life, the whole river ecosystem and the diverse nature of watersides were also recognized as targets for protection.

2.1. Management and control of manufacture, and use of chemicals.

Pollution by toxic/hazardous chemicals such as PCBs and pesticides led people and governments to consider that control of manufacture and use of such chemicals were also necessary for water pollution control. In 1973, the Japanese government enacted the "Law Concerning the Examination and Regulation of Manufacture of Chemical Substances". The law required examination of newly manufactured or registered chemicals before their production or import. In addition, this law also requires the national government to confirm the safety of the existing chemicals. Production and import of chemicals found to be toxic was strictly regulated. When a chemical is concluded to be potentially damaging to human health, its manufacturers are required to investigate and report its toxicity to the government.

Recently, many developed countries have established a new management system for chemicals, called the Pollutants Release and Transfer Register (PRTR). This system requires industries to quantitatively evaluate the discharge of pollutants from their plants to the environment via wastewater, exhaust gas and solid wastes, and then report the results to the local government. Pollutants discharge from non-point sources such as houses and agricultural fields are estimated by the local government from information on sales of the chemicals. The data collected by the local governments are reported to the central government (to the Ministry of International Trade and Industry in Japan).

These data are analyzed by the Environmental Agency to construct a database which is then open to public inspection.

2.2. Minimization of pollutant discharge from production processes.

Most of environmental pollution problems arise from the emission of materials in various forms from industries and human daily life. The development of a material cycling system to minimize the total load to the environment is therefore urgently required. To minimize emissions to the environment, it is first necessary to analyze the material flow and balances in the social system, namely human daily life and production processes. For each local area, the input, output and accumulation of materials for production processes should be determined. The next priority is emission minimization at source. The detailed procedures and approaches for this purpose will be discussed in *Critical Load on Water Body*. Where reduction of emissions from production processes is incomplete the wastes must be reused or recycled. The first priority is reuse/recycling in other units of the same production process, which is called "intra-process recycling". The second is the recycling of wastes in other production processes in the same industry, which is called "trans-process recycling". The recycling of wastes in the processes of other industries, which is called "trans-industry recycling", is the final approach. It should be noted that "trans-industry recycling" may bring about the dissemination of pollutants to other industries. Appropriate treatment or conversion of wastes is essential before they are recycled in other industries.

3. Present situation of water quality

As a result of these countermeasures, water quality, in terms of toxic substances such as heavy metals, cyanide and PCB (Polychlorinated biphenyl), has been remarkably improved in Japan. River water quality in term of organic pollutants (BOD) has also been continually improved. The compliance ratio with respect to the environmental quality standard of BOD for rivers increased from 50% in 1974 to about 80% in 1995. Unfortunately, however, there has been little improvement in lake water quality. The percentage compliance of environmental water quality standard in term of COD for lakes has kept at a low level, ranging from 40% to 45%. Pollution by nutrients (nitrogen and phosphorus) of lake water has become increasingly serious since the early 1990s. More than 33% of all lakes in Japan are considered to be eutrophic. One of the important reasons for the lack of improvement of lake water quality is the diversity and complexity of the sources of pollution. In addition to industrial and domestic wastewaters, agricultural wastewater, livestock wastewater and nutrients derived from aquatic animals and plants are also pollution source. Therefore, conventional measures, such as establishment of effluent standards for industrial and domestic wastewaters and construction of sewerage system, are not sufficient. An overall measure covering all possible pollution sources is required.

Recently, concern about groundwater pollution has also been raised. In the last decade, volatile organochlorines such as trichloroethylene and tetrachloroethylene were detected in groundwater at industrial and urban areas in many developed countries. These organic chemicals are utilized as solvents in many modern industries, laundry firms and domestic usage. Agricultural activity, which utilizes many kinds and large amount of chemicals such as fertilizers and pesticides, is also a potential source of groundwater pollution. Pollution of groundwater by nitrate has become a great environmental issue. There is an urgent need to establish integrated measures for groundwater pollution control.

Glossary

Intra-process recycling:	Reuse/recycling of wastes in other units of the same production process.
Trans-process recycling:	Reuse/recycling of wastes in other production processes in the same industry.
Trans-industry recycling:	Reuse/recycling of wastes in the processes of other industries.

Bibliography

Fujie K. (1997) Zero-Emission Materials Cycle in Production Process and Regional Scale, *Clean Technology*, 3(3), 13-24. [This paper presents the procedures for minimization of pollution discharge from production process and describes issues at a regional scale]

Fujie K. and Hu H.-Y. (2000). Emission minimization and chemicals management in industries and in regional area, In: *Proceeding of the third China-Japan symposium on water environment*, 51-65. [This paper presents approaches for minimization of pollution discharge and chemicals management in industries and regional areas]

Hu H.-Y., Goto N. and Fujie K. (1999). Concepts and methodologies to minimize pollutant discharge for zero-emission production, Water Sci. and Technol., 39(10/11), 6-16. [This paper presents concepts and methodologies for minimization of pollution discharge from production processes]

Japan Environmental Agency (1997) Environment in Japan, Heisei 9 year (In Japanese) [This book overviews the present condition of environmental pollution in Japan]

Odaka M. and Peterson S. A. (2000). *Water Pollution Control Policy and Management: The Japanese Experience*, Gyosei, Japan. [This book overviews the history of water pollution control policy and management in Japan]

Biographical Sketches

Koichi Fujie is a professor in the Department of Ecological Engineering at Toyohashi University of Technology, Japan. He completed his PhD in environmental chemistry and engineering at Tokyo Institute of Technology; his PhD thesis was entitled "Oxygen transfer and power economy characteristics of biological wastewater treatments". Professor Fujie's research and teaching interests are focused on the sustainability of human society supported by industrial activities. He stresses that minimization of resource and energy consumption, with their environment loading, are essential for sustainability. His major research fields are water and wastewater treatment, development of material recycling technology, bioremediation and design of sound material cycle networks.

Hong-Ying HU is a professor and deputy director of the Department of Environmental Science and Engineering at Tsinghua University. He obtained his master and PhD degrees at Yokohama National University, Japan. His major is environmental microbiology and biological engineering. His research area includes kinetic and ecological study on biodegradation of refractory toxic organic chemicals in natural and manmade ecosystems, bacterial community structure and function in ecosystems, biological and ecological technologies for environmental pollution control, and risk assessment and water quality control for wastewater reclamation and reuse.

HYDROLOGIC CYCLE AND WATER USAGE

Koichi Fujie
Department of Ecological Engineering, Toyohashi University of Technology, Japan

Hong-Ying Hu
ESPC, Department of Environmental Science and Engineering, Tsinghua University, China

Keywords: Water stocks, hydrologic cycle, water balance, water usage, water reuse.

Contents

1. Stocks of water on the Earth
2. Hydrologic cycle
3. Water balance and usage
4. Preservation and effective use of water resources
Glossary
Bibliography
Biographical Sketches

Summary

The stocks of water and the hydrologic cycle on the Earth are quantitatively described and the present statuses of water balances and usage in Japan and the USA are compared. The necessity and usefulness of wastewater reuse for saving the limited water resources in the world is also discussed.

1. Stocks of water on the Earth

Water is the origin of life. If there were not water on the Earth, no living things would have come into existence. The life of a living organism is maintained by water contained within its body. Water accounts for about 70% of the body-weight of adult humans, and as much as 80% of that of new-born babies. Generally, the body of a fish is about 75% water; jellyfish, however, contain as much as 96% water. An adult human needs to drink about two liters of water every day. In addition to sustaining the life of every cell, water is also used for many other purposes, such as temperature control, transportation, dissolving materials, washing, maintaining the functions of natural ecosystems, etc. Water is clearly one of the essential resources for human activities.

Water present everywhere on the Earth. It exits in the seas, lakes, ponds, rivers, and under the ground. It also exits in glaciers and mountains as ice and/or permanent snow. Water is also contained in living things, plants, soils and the atmosphere. The water located in lakes, ponds, and rivers is called surface water and that located under the ground is called groundwater or subsurface water. Fresh surface water and groundwater are major water resources for human activities. The stocks and location of water on the Earth are shown in Table 1.

Location	Amount (10^3 km^3)	Percentage*
Oceans	1,357,000	96.26
Subsurface	25,700	1.82
Rivers	1.4	0.00099
Freshwater lakes	91	0.0065
Saline lakes	85	0.0060
Glaciers (ice and permanent snow)	26,410	1.87
Soil	80	0.0056
Atmosphere	13	0.00092
Living biomass	1.2	0.000085
Total	**1,409,607**	

*The approximate values (the sum of the data does not equal to 100%)

Table 1. Stocks of water on the Earth
Data from Naganuma, 1978 and Masters, 1997

The water in the seas is estimated to be 1 357 000x10^3 km^3. This is based on the total area of the World Ocean being about 361x10^6 km^2, and the mean depth about 3800 m. More than 96% of the world's water is located in the seas. The amount of groundwater is about 25 700 x 10^3 km^3, which accounts for about 1.82% of the total stock of water in the world, and half the total fresh water. The water in fresh lakes, ponds and rivers is 92 x10^3 km^3, which is only 0.007% of total water and 0.18% of fresh water. The amount of water in the forms of ice and permanent snow is about 26 410 x 10^3 km^3, which is nearly equal to the amount of groundwater.

The total area of glaciers and ice sheets in the world is about 16 x 10^6 km^2 (about 10% of the land area), and 90% of that is located in Antarctica. The amount of water in soils is about 80 x 10^3 km3, which is similar to the volume stored in fresh lakes. The water contained in the atmosphere is only 13 x 10^3 km^3, or 0.0009% of total water. In addition to water vapor, it includes various natural phenomena such as clouds, rain and snow.

2. Hydrologic cycle

The location, state and/or form of materials on the Earth are dynamically changing all the time inside the Earth spheres. Water for human activity is provided by a natural cycle, driven by precipitation (rain and snowfall). Figure 1 shows a highly simplified illustration of the natural cycle of water.

The total annual precipitation on the Earth is about 543 x 10^3 km^3. Roughly 80% of that (431 x 10^3 km^3) falls into the ocean, while the remaining 20% (112 x 10^3 km^3) falls on the land. The rainfall and snowfall on the land become surface water and finally flow into the seas through rivers. A part of the surface water may sink into the subsurface of the Earth and become groundwater, which will also flow into the seas, usually after a long delay. Part of the surface water and groundwater are withdrawn for human daily life and industrial and economic activities on their way to the seas. After use this water (called wastewater) is then returned to rivers and lakes, generally after appropriate treatment. The discharge of wastewater without appropriate treatment causes serious pollution problems for the aquatic environment.

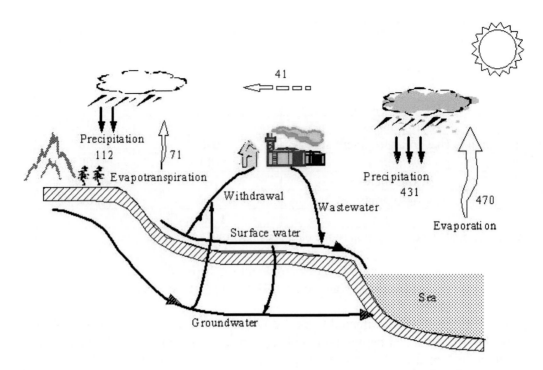

Figure 1. Illustration of the hydrologic cycle on the Earth (units: 10^3 km^3 per year)
(Note: Illustration modified from Tanbo, 1995, Data from Takeda, 1994 and Master, 1997)

The precipitation on the Earth is supplied by the evaporation of water from the seas and the land. The sun provides the energy needed for this evaporation. About 470 x 10^3 km^3 of water is evaporated from the seas to the atmosphere each year. Water is also transferred from the land to the atmosphere by evaporation from the wet surface of land, lakes and rivers and by transpiration from the leaves of plants. The combination of evaporation and transpiration is called evapotranspiration. The precipitation of water onto the land is about 112 x 10^3 km^3, which is less than the evapotranspiration from it. The difference between the precipitation and evapotranspiration is transferred as vapor from the seas to the land through the atmosphere. It can be imaged that the sea is a huge factory producing fresh water. The fresh water made in the factory is pumped to the land by the movement of the atmosphere. The water pumped to the land then returns to its hometown of the seas. This macro-cycle of water on the Earth is called the hydrologic cycle.

The average retention time of water in the atmosphere is estimated to be about seven to nine days. In other words, the water in the atmosphere is totally replaced every seven to nine days. The retention time of groundwater and seawater are about one to two years, and about 3000 years, respectively.

3. Water balance and usage

Water balance and usage vary between countries and areas. The balances and usages of fresh water in Japan and the USA are shown in Figure 2.

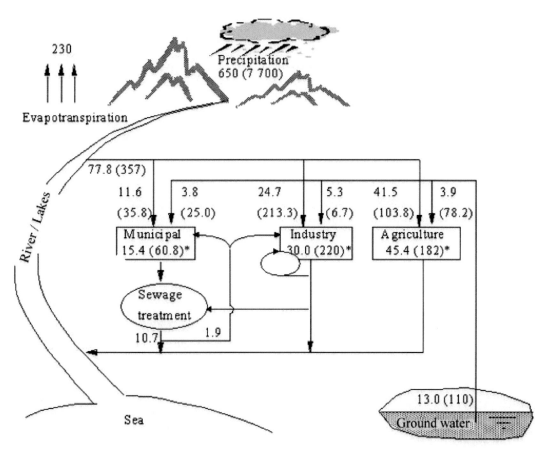

Figure 2. Water balances and water usage in Japan and the United States
(data for the United State are shown in parenthesis)
*(Note: Data from WRI, UNEP, UNDP, and The World Bank, 1998, Master, 1997, and
Goto, Hu and Fujie, 1999; Industry includes power plant and mining; Units: km^3 per
year; *: Amount of total withdrawal)*

The total precipitation in Japan is about 650 km^3 per year. The annual precipitation per capita is about 5300 m^3, which is only one fifth of the world average value of 27 000 m^3. The total annual withdrawal of fresh water was about 90.8 km^3 in Japan in 1990 (annual withdrawal per capita: 734 m^3). About 86% of that is collected from rivers and lakes and the remaining 14% is supplied by groundwater. Seventeen percent (15.4 km^3) of the total withdrawal is for municipal uses, another 30% (30.0 km^3) is for industrial uses and about 50% is for agricultural uses. The total amount of sewage generated in Japan was about 10.7 km^3 in 1994, and about 1.7 km^3 of sewage effluent was reused after appropriate treatment.

The annual precipitation of the USA is about 7700 km^3 in total, and 28 000 m^3 per capita. The per capita precipitation in the USA is therefore about five times that in Japan. The total withdrawal of fresh water is about 467 km^3 each year, 76% of which is collected from rivers and lakes and the other 24% from groundwater. The total withdrawal of fresh water per capita in the USA is 1840 m^3, which is 2.5 times that in Japan. About 47% of the total withdrawal of fresh water in the USA (220 km^3) is for industrial use (including power plants and mining), another 13% (60.8 km^3) is for municipal use and the remaining 40% (182 km³) is for agriculture. The fresh water consumed at power plants accounts for about 85% of the total consumption by industry in USA. In Japan, however, only 5 to 6% of water for industrial use is consumed by power plants. In addition, the municipal use of fresh water in the USA per capita is calculated to be about 610 liters per day, which is almost twice that in Japan (340 liters per day per capita in 1990).

4. Preservation and effective use of water resources

Global average annual per capita precipitation is about 27 000 m³. Per capita availability of water differs greatly between countries and even between different parts of a country; the geographic distribution of fresh water is not well matched to the distribution of people on the Earth. Many countries and area are facing serious problems of water shortage. How to effectively use the limited resource of fresh water is one of the critical issues of the twenty-first century. Reuse of wastewater is a useful approach for saving water resources and reducing pollution of natural water bodies.

Wastewater reuse was promoted from the 1960s in Japan. As shown in Figure 3, the average ratio of wastewater reuse linearly increased from 1965, reached 74% by the end of the 1970s and about 77% in the 1990s.

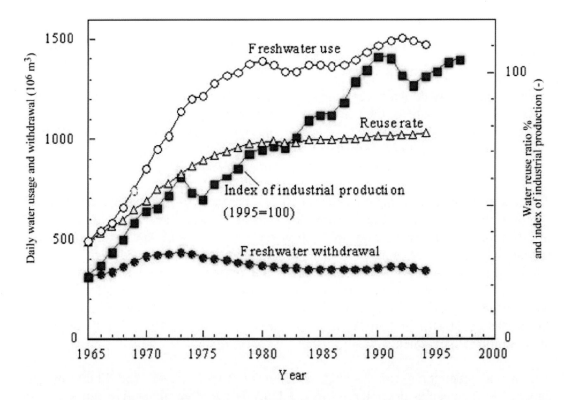

Figure 3. Industrial freshwater use, withdrawal and reuse in Japan.
(Data from National Land Agency Japan, 1997)

As a result, fresh water withdrawal started to decrease from 1970 although water usage and industrial production continuously increased. It should be noted that the reuse ratio of wastewater is very different between industries. For example, the reuse ratios in steel and iron industries, oil refinery plants, chemical industries and automobile industries are as high as 80 to 90%. Unfortunately, however, the reuse ratios in the pulp (about 40%), food (about 38%) and textile (about 20%) industries are still relatively low. Increased reuse of wastewater in those industries should be encouraged by statutory means, if voluntary change and financial incentives are insufficient.

Glossary

Evapotranspiration:	Loss of water in an area through evaporation from the wet surface of land, lakes and rivers and transpiration from the leaves of plants.
Hydrologic cycle:	The natural cycle of water in the Earth's spheres. The location and form of water on the Earth are dynamically changing all the time.

Bibliography

Fujita S. and Sono K. (1990). Water and lives (living science of water). 1-5. Tokyo: Makishoten Co., Ltd. [In Japanese]. [This book discusses the function and importance of water for living things].

Goto N, Hu H.-Y. and Fujie K. (1999). Energy consumption and material flow in Japanese sewage treatment plant, *Proceedings of 7th IAWQ Asia-Pacific Regional Conference*, Vol.1, p.59-64. [This paper overviews the material flow and amount of energy consumed in Japanese sewage treatment plants].

Master G.M. (1997). Introduction to environmental engineering and science (2nd ed.). 164-170. New Jersey: Prentice Hall, Inc. [This book gives a brief and systematic introduction on principles of environmental engineering and science].

Naganuma N. (1978). Water sphere of the Earth. Ocean and land waters, (ed. M. Hoshino), 1-5. Tokyo: Tokai University Press. [In Japanese]. [This book gives a brief introduction to the water resources of the Earth].

National Land Agency Japan (1997). Water resources in Japan -1997. 28-51. Tokyo: Printing Bureau the Ministry of finance Japan. [In Japanese]. [This book overviews water resources in Japan].

Takeda T., Ageta Y., Yasuda N. and Fujiyoshi Y. (1994). Meteorology of water cycle. 1-4. Tokyo: University of Tokyo Press. [In Japanese]. [This book gives a systematic introduction to the water cycle on the Earth].

Tanbo N. and Ogasawara K. (1995). Water purification. 1-10. Tokyo: Gihodo Shuppann Co., Ltd. [In Japanese] [This book gives a systematic introduction to the water cycle on the Earth].

Wada H. (1992). Water recycling (Basic). 2-4. Tokyo: Chijinshokan Co., Ltd. [In Japanese]. [This book gives a systematic introduction to wastewater reclamation and reuse technologies].

WRI, UNEP, UNDP, and The World Bank (1998). World resources 1998-1999: A guide to the global environment. 302-306. Tokyo: Chuohoki Publishers. [In Japanese] [This book overviews the present condition of resources in the world].

Biographical Sketches

Koichi Fujie is a professor in the Department of Ecological Engineering at Toyohashi University of Technology, Japan. He completed his PhD in environmental chemistry and engineering at Tokyo Institute of Technology; his PhD thesis was entitled "Oxygen transfer and power economy characteristics of biological wastewater treatments". Professor Fujie's research and teaching interests are focused on the sustainability of human society supported by industrial activities. He stresses that minimization of resource and energy consumption, with their environment loading, are essential for sustainability. His major research fields are water and wastewater treatment, development of material recycling technology, bioremediation and design of sound material cycle networks.

Hong-Ying HU is a professor and deputy director of the Department of Environmental Science and Engineering at Tsinghua University. He obtained his master and PhD degrees at Yokohama National University, Japan. His major is environmental microbiology and biological engineering. His research area includes kinetic and ecological study on biodegradation of refractory toxic organic chemicals in natural and manmade ecosystems, bacterial community structure and function in ecosystems, biological and ecological technologies for environmental pollution control, and risk assessment and water quality control for wastewater reclamation and reuse.

MINIMIZING LOADS ON WATER BODIES

Koichi Fujie

Department of Ecological Engineering, Toyohashi University of Technology, Japan

Hong-Ying Hu

ESPC, Department of Environmental Science and Engineering, Tsinghua University, China

Keywords: Self-purification, kinetics of waste removal, pollutant minimization, negative flow sheet, on site treatment, closed process.

Contents

1. Self-purification of natural water bodies
2. Kinetics of self-purification in natural water bodies
3. Minimization of pollutants loading to natural water bodies
Glossary
Bibliography
Biographical Sketches

Summary

The concept and mechanisms of self-purification of natural water bodies are described in this article. Approaches for minimizing pollutant loading to natural water bodies are proposed and the usefulness of these approaches is discussed.

1. Self-purification of natural water bodies

It is a well-known fact that the wastes/pollutants discharged into natural water bodies such as rivers; lakes and the seas disappear slowly with time. The removal of pollutants from a water body without any artificial controls is called self-purification, or natural purification. The mechanisms of self-purification of water bodies can be divided into three groups: physical processes, chemical processes and biological processes.

Physical processes contributing to the removal of pollutants from a natural water body include dilution/mixing by inflow of unpolluted water into the water body, diffusion of pollutants in water, and precipitation/filtration of the pollutants to the sediment. The volatilization/vaporization of volatile pollutants from water to the atmosphere will also result in a decrease of pollutants in the water.

Chemical processes related to the removal of pollutants from a water body are oxidation by oxidants such as ultraviolet, ozone and oxygen, reduction by reductants, and neutralization. The biological processes include degradation/transformation of organic pollutants by bacteria under aerobic or anaerobic conditions, and nitrification and denitrification of ammonia and nitrate, respectively. Biological processes play the most important role among the mechanisms of self-purification in natural water bodies. The biological removal of pollutants from a natural water body is usually called "true self-purification" and the total purification by physical, chemical and biological processes is called "apparent self-purification". Factors affecting the capacity of self-purification of natural water bodies will be discussed in the next section of this article.

2. Kinetics of self-purification in natural water bodies

The amount/concentration of organic matter present in water is one of the most important indices of water quality. *Biochemical oxygen demand* (BOD) is often used to measure the amount of organic matter in water. BOD is defined as the amount of oxygen required for the biological degradation of organic matter in water under aerobic conditions. This can be determined under laboratory conditions. The value of BOD depends upon not only on the concentration of organic matter but also on the temperature and time of incubation, as well as the concentration and type of bacteria used. BOD is usually determined under a standardized temperature and time of incubation, for example 20 C and 5 days.

The capacity for self-purification of a water body may be quantitatively evaluated by its removal rate of organic pollutants (usually expressed as BOD). Generally, the reduction rate of BOD in a water body is proportional to the concentration of BOD residing in water (Equation 1).

$$-dC/dt = kC \qquad (1)$$

Where, C (unit: mg L^{-1}) is BOD at time t, and k (unit: day^{-1}) the total removal rate constant of BOD. By dividing the total removal of BOD into two parts: physical and chemical removal and biological removal, k can be expressed by Equation 2, and Equation 1 can be rewritten as Equation 3.

$$k = k'_b X + k_{pc} = k_b + k_{pc} \qquad (2)$$

$$-dC/dt = (k_b + k_{pc})C \qquad (3)$$

The symbols of k_b' (unit: L mg^{-1} day^{-1}) and k_b (=k_b' X, unit: day^{-1}) are defined here as the specific and overall biodegradation rate constants of BOD, respectively. The X (unit: mg L^{-1}) shows the concentration of active biomass in water and k_{pc} (unit: day^{-1}) expresses the physical and chemical removal rate constant of BOD. The specific biodegradation rate constant of BOD (k_b') is controlled by the kind of organic matter, environmental conditions, and the community structure of bacteria inhabiting the water, and other factors. The overall biodegradation rate constant of BOD (k_b) is proportional to the concentration of active biomass (X) in water. Compared to k_b', the term of k_b is easy for practical use because X is difficult to measure. The term k_b', however, is useful to understand of the affecting factors of BOD degradation in a water body.

By integrating Equation 3, BOD at any time, t, is given as Equation 4 or Equation 5.

$$C = C_0 e^{-(k_b + k_{pc})t} \qquad (4)$$

$$C=C_0 10^{-(K_b + K_{pc})t} \qquad (5)$$

$$(K_b + K_{pc}) = (k_b + k_{pc})/ln10 = 0.43(k_b + k_{pc}) \qquad (6)$$

It is very difficult to determine the values of K_b and K_{pc} individually for natural water bodies. Therefore, a combined value of K_b and K_{pc}, namely (K_b+K_{pc}), is usually measured.

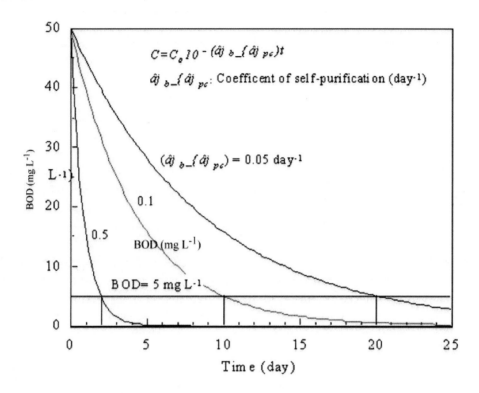

Figure 1. Simulation of BOD degradation in natural water bodies with different capacity of self-purification. BOD at t = 0 is C_0 = 50 mg L^{-1}.

The phenomenon of self-purification in a water body can be illustrated with Figure 1. Figure 1 shows the simulation results of BOD removal in water bodies with different values of (K_b+K_{pc}). The initial BOD is 50 mg L^{-1} for all the cases. The times needed to reduce the BOD from 50 to 5 mg L^{-1} are 20, 10 and 2 days for the cases of (K_b+K_{pc})=0.05, 0.1 and 0.5 day^{-1}, respectively. It is easy to understand that the term of (K_b+K_{pc}) can be used to quantify the capacity of self-purification of a water body. A large value of (K_b+K_{pc}) shows a higher capacity of self-purification. So the term of (K_b+K_{pc}) is usually called the coefficient of self-purification.

Table 1 shows examples of the determined value of (K_b+K_{pc}) for some rivers in Japan. The value of (K_b+K_{pc}) varies from 0.1 to 2 day^{-1}. The factors affecting the capacity of self-purification of a river are depth of water, flow rate, temperature, etc. It has been reported by Kunimats (1990) that lower water temperature gave larger value of (K_b+K_{pc}), and higher flow rate of water decreased the value of (K_b+K_{pc}). It is well known that the specific biodegradation rate constant of BOD decreases with the decrease in temperature just as expressed by Arrhenius's Law. The larger value of (K_b+K_{pc}) at lower temperature may be due to higher concentration of river biomass caused by an increase in density of biofilm attached to solid surfaces. Hu *et al.* (1994) has reported an increasing tendency of microbial concentration in a bioreactor for wastewater treatment under lower temperature. On the other hand, higher flow rate of water may reduce the concentration of biomass in a river by restraining the growth of bacteria on the solid surfaces of the river-bed, resulting in a lower value of (K_b+K_{pc}).

Rivers	$K_b + K_{pc}(\text{day}^{-1})$
Tama river	0.1 - 1.88
Tikuma river	1.2 - 2.4
Matuura river	0.12 - 18
Chikugo river	0.11 - 0.17
Sagami river	0.17 - 0.32

Table1. Coefficients of self-purification ($K_b + K_{pc}$) for some rivers in Japan.

3. Minimization of pollutant loading to natural water bodies

As described above, natural water bodies can purify/assimilate a certain quantity of wastes before becoming polluted. A water body will be polluted when the pollutant discharged into it excedes its capacity of self-purification. Improving the capacity of self-purification and minimizing the loading of pollutants are valuable approaches to pollution prevention. Recently, the pollution of water bodies by refractory and toxic chemicals, which are hard to assimilate in a natural water body, has become more and more serious. So, there is a limit to improving the capacity of self-purification. Minimizing the pollutant loading becomes more and more important.

3.1. Minimizing discharge of pollutants to water bodies

Methods of minimizing the discharge of pollutants from production processes are categorized as follows:

Reduction of pollutant discharge

- Refine or replace the raw and secondary materials
- Refine the unit process to increase yield of product
- Replace the process with alternatives
- On site treatment and closed process

Appropriate selection and operation of treatment processes

- Performance evaluation of existing treatment processes
- Improvement of existing treatment processes
- Appropriate selection/combination of existing treatment methods

Application of newly developed technologies for waste treatments, such as:

- Membrane filtration technologies for removal of pollutants
- Anaerobic digestion for high strength wastewater treatment
- Biotechnology for degradation of refractory organic matter, etc.

The first approach is the reduction of pollutants from factories. The first priority is to reduce the discharge of pollutants from each unit process in a factory. This can be put into practice by increasing the yields of products. Replacing and/or refining the present unit processes in a factory are useful means of increasing product yield, i.e. to reduce the discharge of wastes from the production process, per unit of production. Raw and secondary materials used in the production process should also be selected for this purpose. Raw materials with high quality or high content help to reduce production of waste in the production process. The use of biodegradable raw and secondary materials may be useful in improving the performance of the wastewater treatment process. A production process continuously discharging huge amount of pollutants may require installation of an alternative production process based on a new concept. Establishing a production process incorporating on-site treatment and recycling f wastes, i.e. a closed system, is the second priority. Treatment of waste materials and wastewaters should be incorporated into the production process for easy recycling of treated waters and materials.

The second approach is the optimal selection and operation of a wastewater treatment process. Appropriate combination of various wastewater treatment methods should be considered to give high quality effluent with less energy consumption. This concept can be also applied for the treatment of waste materials. Treated materials should be recycled or reused for production in the same factory or in another factory, or even another industry.

The third approach is to develop advanced technologies for the treatment of waste materials and waters in accordance with the above-mentioned requirements, i.e. reduction of energy consumption and high quality effluent.

3.2. Procedures to minimize pollution discharge

Analysis of pollution source and pollutant loading: First of all, the source, quality and quantity of the pollutants discharged from each production process should be determined before considering the reduction of the discharge of pollutants. A mass balance focused on wastes (i.e. non-products) in a production process (called a 'negative flow sheet') is very useful in understanding the situation of the production process and identification of the pollution sources, i.e. the quantity and characteristics of emissions from the production process.

In some cases, the discharge of pollutants may be simply reduced by changing the sealant and by refining or changing the raw materials. Optimization of reaction schemes and operating conditions will bring about an increase in the conversion of raw materials to obtain a high yield of products. Increase in the yield of products brings about a reduction of emissions. The production process may need to be replaced by an alternative process to reduce or eliminate the discharge of pollutants. For example, super-critical carbon dioxide (SCCD) may be a preferable choice for replacing the toxic organic solvents used in industrial extraction processes. SCCD is a non-polar solvent with non-toxic and non-flammable characteristics. Organic solvent extraction may be replaced by SCCD extraction. Electrolysis using a mercury electrode to produce sodium hydroxide and chlorine from sodium chloride has been successfully replaced by electrolysis using ion exchange membranes, thus eliminating wastes containing mercury.

Improvement of production process. The procedures to improve the production process in chemical industry are shown in Table 2.

Improvement	Examples
Management and facilities	Refine and improve the sealant, tightness and raw materials used
Reaction	Change solvent; Increase the conversion, yield and product concentration
Operation	Replace batch operation with continuous operation
Cleaning	Improve the method and interval; Cleaning without water

Table 2. Procedures to reduce pollutant discharge from production processes in the chemical industry.

An example of the reduction in wastewater discharge from a polymerization process in Japan is shown in Table 3.

	Jan 1965	Jan 1973	Nov 1973	Feb 1974
Wastewater (m^3/t-polymer)	36.2	17.0	8.8	0.2
Countermeasure		Increase in polymer concentration in liquid.	Wastewater recycling from separation process	Wastewater recycling from catalyst washing and polymer drying processes

Table 3. Example of reduction in the wastewater from a polymerization process
(Modified from Saeki, 1979)

In 1965 more than 36 m^3 of wastewater was discharged from the polymerization process to produce one ton of polymers. The increased polymer concentration in the reactor was effective at cutting 50% of the wastewater produced. Wastewater recycling from the polymer separation process after appropriate treatment further reduced the residual wastewater by another 50%. The recycling of wastewater from the catalyst washing and the polymer drying processes also brought about a noticeable reduction in wastewater discharge, so the utilization of water in the whole polymerization process saw further significant reduction. Moreover, gas phase polymerization or liquid phase bulk polymerization processes can almost entirely eliminate wastewaters.

On-site treatment and closed processes. Wastewater treatment involves a separation process for mixed pollutants with different characteristics. Mixing of wastewaters from different sources makes the separation more difficult due to their different characteristics, example.g. different solubility, polarity, biological or chemical degradability, coagulation characteristics, molecular size, etc. On-site treatment of wastewater with recycling of treated wasters, as shown in Figure 2 is an alternative for treatment of wastewater. In this process, wastewater is treated separately without mixing.

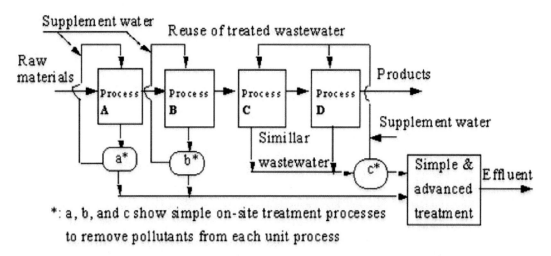

Figure 2. Possible procedures to reduce the discharge of wastewaters from production processes by incorporating a recycling system with on-site wastewater treatment
(Modified from Saeki, 1979).

Separation of the pollutants contained in one kind of wastewater is easy compared to that in mixed wastewater. Pollutants can be removed using a simple treatment process for wastewater and the treated water can be recycled to the unit process from which the wastewater is discharged. Thus a closed production process can be established from the viewpoint of water utilization.

Glossary

Biochemical Oxygen Demand (BOD):	The amount of oxygen required for the biological degradation of organic matter in water under aerobic conditions.
Denitrification:	The process by which bacteria transform nitrates to nitrogen gas.
Negative flow sheet:	A mass balance focused on wastes (i.e. non-products) in a production process. This is very useful for identification of the pollution sources, i.e. the quantity and characteristics of emissions from the production process.
Nitrification:	The process by which bacteria transform ammonia to nitrates.
Self-purification:	The phenomenon that water bodies can clean themselves of polluting substances.

Symbols

C:	BOD in water (mg L^{-1})
T:	Time (day)
X :	Concentration of active biomass in water (mg L^{-1})
k :	Total removal rate constant of BOD (day^{-1})
k_b :	Overall biodegradation rate constants of BOD (L mg^{-1} day^{-1})
k_b':	The specific biodegradation rate constants of BOD (L mg^{-1} day^{-1})
k_p:	Physical and chemical removal rate constant of BOD (day^{-1})

Bibliography

Hu H.-Y., Goto N. and Fujie K. (1999). Concepts and methodologies to minimize pollutant discharge for zero-emission production, Water Sci. and Technol., 39(10/11), 6-16. [This paper presents concepts and methodologies for minimization of pollution discharge from production process]

Hu H.-Y., Fujie K. and Urano K. (1994). Effect of temperature on the reaction rate of bacteria inhabiting the aerobic microbial film for wastewater treatment, *J. Ferment. Bioeng.,* 78(1), 100-104. [This paper discusses the effect of temperature on the amount and activity of biomass in a biofilter]

Kunimatsu T. and Muraoka K. (1990) Modeling analyses of river pollution, 101-108. Tokyo: Gihodo Shuppann Co., Ltd. [In Japanese]. [This book gives a systematic introduction to self-purification, sources of river pollution, and methods for pollution load estimation]

Saeki Y. (1979). Closed system of chemical production processes. Tokyo: Kogyo Chosakai Publishing Co. Ltd. [In Japanese]. [This book introduces approaches and practices for minimization of pollutant discharge from production processes]

Biographical Sketches

Koichi Fujie is a professor in the Department of Ecological Engineering at Toyohashi University of Technology, Japan. He completed his PhD in environmental chemistry and engineering at Tokyo Institute of Technology; his PhD thesis was entitled "Oxygen transfer and power economy characteristics of biological wastewater treatments". Professor Fujie's research and teaching interests are focused on the sustainability of human society supported by industrial activities. He stresses that minimization of resource and energy consumption, with their environment loading, are essential for sustainability. His major research fields are water and wastewater treatment, development of material recycling technology, bioremediation and design of sound material cycle networks.

Hong-Ying HU is a professor and deputy director of the Department of Environmental Science and Engineering at Tsinghua University. He obtained his master and PhD degrees at Yokohama National University, Japan. His major is environmental microbiology and biological engineering. His research area includes kinetic and ecological study on biodegradation of refractory toxic organic chemicals in natural and manmade ecosystems, bacterial community structure and function in ecosystems, biological and ecological technologies for environmental pollution control, and risk assessment and water quality control for wastewater reclamation and reuse.

GROUNDWATER DEGRADATION BY HUMAN ACTIVITIES

Yoshiteru Tsuchiya
Department of Applied Chemistry, Faculty of Engineering, Kogakuin University, Japan

Keywords: groundwater, human activities, chemical substances, metals, organic chlorinated chemicals, agricultural chemicals, chromium, lead, cadmium, mercury, arsenic, pesticide, PPCPs

Contents

1. Introduction
2. Causative materials to contaminating groundwater
3. Metals and inorganic compounds
4. Chlorinated organic compounds
5. Agricultural chemicals
6. Pharmaceutical and personal care products (PPCPs)
7. Environmental pollution by the final disposal site leachate.
8. Environmental standard item and detection status of the groundwater in Japan
Glossary
Bibliography
Biographical Sketch

Summary

Groundwater shows generally settled and good water quality; however, its quality may deteriorate when contaminated with inorganic and organic chemicals by human activity. Under seepage from industrial liquid waste, drainage from livestock facilities, domestic drainage and mine drainage are considered causes of groundwater pollution. In industrial waste, they are metals such as chromium, cadmium, and nickel, which come from metal plating factory drainage, and volatile organic compounds (VOCs) such as trichloroethylene, and tetrachloroethylene used for degreasing and metal cleaning. In the discharge from livestock facilities can be detected microorganisms, organic matter, nitrogen compounds and antibiotics used for livestock. From domestic drainage are found organic matter and detergents, and a micro amount of pharmaceuticals. Pesticides for agricultural use are also often detected in groundwater

1. Introduction

Groundwater is widely used for various purposes such as domestic and industrial needs. Its origin is falling rain and snow reaching the surface of the earth and then permeating underground to form a water zone. Groundwater consequently contains a solution of inorganic components found in the crust of the earth, and undesirable pathogenic microorganisms are generally removed through filtration during the infiltration process. Since ancient times, humans have utilized groundwater as drinking water, as domestic water for cooking and washing, and more recently, for agriculture, fisheries, and industry. In ancient ruins excavated in various parts of the world, large wells have been found in the center of ancient cities.

Since the industrial revolution, as mining and manufacturing developed rapidly, artificially synthesized chemical substances have been allowed to permeate the land. Such chemicals are carried underground where they degrade the condition of the groundwater. As scientific technologies and precision machine industries have developed, the problem of groundwater contamination by artificially synthesized organic compounds has become more widespread around the world.

Of the various substances dissolved from crust components during the process of rainwater infiltration underground, such substances as arsenic, nitrate ions and nitrite ions have a harmful influence on human health; however, those substances are never produced by ordinary human activities.

Major threats to human health occur when ions of cyanide, chromium and lead, which are particularly used in electro-plating plants, are allowed to contaminate groundwater. Recently, it was reported that groundwater had been contaminated by chlorinated organic chemicals (trichloroethylene and tetrachloroethylene) in Silicon Valley, California, USA. Great efforts of various kinds are required to prevent water contamination, particularly groundwater, which is essential for human life as the principal source of clean fresh water.

2. Causative materials contaminating groundwater

Under seepage from industrial liquid waste, drainage from livestock facilities, daily life drainage and mine drainage are considered activities during which material from the surface permeates the groundwater and becomes a cause of pollution.

As mining developed around the world, a huge amount of groundwater was removed and put to use. A rapid increase in the population has also raised the need for water for daily life, and huge amounts of groundwater have been used for this purpose. If groundwater is extracted at a rate greater than the natural recharge rate, aquifers will become depleted and may become unusable. Furthermore, extraction may cause subsidence of the ground surface over the aquifer. In addition, developments in mining and manufacturing industries, and the modernization of agriculture have contaminated groundwater due to the presence of harmful chemical substances. Such substances as metals, cyanides, pesticides and chlorinated organic chemicals are known to be causative substances.

As a result of industrial liquid waste, there are metals such as chromium, cadmium and nickel, and cyanogen compounds which come from metal-plating factory drainage, and volatile organic compounds used for degreasing and metal cleaning. In the discharge from livestock facilities can be detected microorganisms, organic matter and nitrogen compounds. From domestic drainage are found organic matter and detergents, and in mine drainage are often detected metals and inorganic materials. In addition, nitrate ions are found, which derive from nitrogen fertilizer and pesticides around the farm.

3. Metals and inorganic compounds

3.1 Chromium

Chromium is used for surface metal coating in electro-plating plants. It is first processed by chemical precipitation or adsorption methods using ion exchange resin and charcoal, and then it is drained. During the washing process, if water leakage or illegal liquid waste dumping occurs, chromium-contaminated water may infiltrate the ground, and cause groundwater contamination. Furthermore, since the metal-coating process usually requires cyanide, such as in potassium cyanide, this can cause groundwater contamination in the same manner as chromium. Both chromium (VI) and chromium (III) are known as stable chemical forms. Of the two substances, chronic exposure to chromium (VI) has the potential to cause pulmonary cancer in humans, but its exposure route is mainly by inhalation through the respiratory tract. Such symptoms as vomiting, diarrhea and stomach pains are known to be attributable to chromium (VI) poisoning, and nephropathy is sometimes reported. Another example of a situation that can lead to groundwater contamination is inadequate treatment of waste from puddling plants. If this is allowed to infiltrate groundwater, the metals it contains can present a serious threat to human health. The most dangerous are lead, cadmium, mercury and selenium. Chronic exposure to lead inhibits the generation of hemoglobin in blood, which causes anemia. It may also become a cause of nephropathy due to induction of the abnormal metabolism of sodium in the kidneys. Chronic exposure to cadmium also causes nephropathy with such symptoms as proteinuria and glycosuria. Inorganic mercury is easily accumulated in the kidney, causing damage to epidermal cells of the proximal tubules.

3.2 Arsenic

Country/Region	Concentration range ($\mu g \ L^{-1}$)	Area (km^2)	Population exposed
Bangladesh	<0.5-2500	150,000	Ca. 3×10^7
West Bengal	<10-3200	23,000	6×10^6
Taiwan	10-1820	4000	10^5
Inner Mongolia	<1-2400	4300	?
Xinjiang	40-750	38,000	500
Hungary	<2-176	110,000	29,000
Argentina	<1-5300	10^6	2×10^6
Northern Chile	100-1000	125,000	500,000
South-west USA	<1-2600	5000	3×10^5
Mexico	8-620	32,000	4×10^4

Table 1: Naturally-occurring As problems in world groundwater (UN2001)

Arsenic is used widely such as for semiconductor material, pigments, and pesticides, and it is also widely distributed in natural geology. Recently, it was reported that in Asian regions such as the Bengal district of India, China and Bangladesh, arsenic, as a natural component of the Earth's crust, is dissolved in groundwater, causing serious effects on the health of local people, particularly where there is no other water supply. Long-term uptake of arsenic in water causes dermatocarcinoma.

The results, which the United Nations investigated in 2001, are shown in Table -1. It is being detected in the concentration range of μg L^{-1} ~ mg L^{-1} levels from groundwater in many countries.

The main cause is from geology and it causes arsenopyrites (FeAsS), and arsenic seems to dissolve in the groundwater. However, one cause of the arsenic pollution of groundwater is that deeper groundwater is used which contains arsenic from drought. The cause of the drought is suspected as being a result of climatic change by global warming.

3.2.1. Chemical form and environment dynamic phase of the arsenic.

Arsenic is a metalloid element which can be either a metal or nonmetal, and many organic arsenic compounds which show different chemical property to inorganic are known to have multiple redox states, and in a charcoal element to form a stable covalent bond. Although arsenic generally exists stably under oxidative conditions near neutrality in the condition of   pentavalent valence, it easily reduces under reductive conditions and with biological action, and it is alkylated, and changes to a chemical compound of three valences which show high reactivity and toxicity. Arsine and these alkyl derivatives are volatile at ordinary temperature and show high toxicity. Arsenic acids, methanearsonic acid salt, cacodylic acid, are used as pesticides such as in wood preservative, herbicide, and insecticide. Sprayed arsenic compound moves through the environment, influenced by the redox state, pH, biological activity and others. It returns to the ground in rainwater as atmospheric arsine, which is oxidized to pentavalene in air, and it may permeate underground.

3.3 Cyanide

Cyanide dissolved in water causes a change in pH of the solution and is also diffused as volatile hydrogen cyanide. When it reaches the groundwater, it is gradually volatilized, and eventually decomposes. When cyanide is taken orally, under acid pH conditions in the stomach, hydrogen cyanide is liberated. Some of this is then taken into the lungs, where it inhibits the effect of cytochrome oxidase in the electron transport system to terminate cellular respiration.

3.4 Nitrate and nitrite

Nitrate nitrogen (NO_3-N) is detected frequently in the groundwater at high density as well as as a volatile organochlorine compound. In humans, nitrate nitrogen is reduced to nitrite nitrogen, and causes methemoglobinemia (cyanosis, suffocation symptoms).

As a supply source in groundwater, it infiltrates rainfall, agriculture materials (organic and inorganic fertilizers, plant residue), livestock drainage, domestic drainage and industrial liquid waste.

- Rainfall: Although nitrogen oxides exist in the atmosphere, and therefore, nitrate nitrogen is contained in rainfall, its load is limited, because the concentration decreases with the increase in the amount of rainfall.
- Agriculture: The nitrogen fertilizers used for farmland permeate underground, and become a problem.

The nitrogen fertilizers used in paddy fields have limited use, and the effect on the groundwater is limited from absorption to the crop. 1t ha^{-1} per year may be used in a plowed field crop for eclipse leaf vegetables, and a considerable part may permeate underground. For example, the groundwater of farmland, which uses 500kg h^{-1} of nitrogen fertilizer in one year, has 13 times higher nitric acid than groundwater without fertilizer.

- Waste water treatment: When domestic drainage and industrial liquid waste infiltrate soil, high nitrate nitrogen is often found in the groundwater. In particular, nitrate nitrogen concentration rises with the soil infiltration of septic waste discharge.

In general, the discharge contains organic nitrogen (Org-N), ammonia nitrogen (NH$_4$-N), nitrite nitrogen (NO$_2$-N), and nitrate nitrogen, and these are oxidized in the soil to nitrate nitrogen. The result of dealing with the discharge of septic waste in a soil infiltration disposing facility is shown in Table 2. Nitrate nitrogen is increased remarkably in the groundwater, while NH$_4$-N and org-N are decreased by soil infiltration.

	Effluent (mg L^{-1})	Seepage water (mg L^{-1})	Removal rate (%)
BOD	53.1	0.6	99
NH$_4$-N	125	8.6	93
NO$_2$-N	0.2	0.89	-
NO$_3$-N	0.22	87	-
Org-N	6.84	0.83	94
Total nitrogen	132	95.8	27
Total phosphorous	9.75	0.03	100

Table 2: Water qualities of effluent and soil infiltration water

Partly to minimize the exposure to metals which cause damage to human health, the WHO maintains guidelines for the allowable concentrations of contaminants in drinking water. These guidelines include both organic and inorganic compounds, some of which are naturally present in the Earth's crust. These guidelines are used when national drinking water standards are established.

4. Chlorinated organic compounds

Since the nineteenth century, humans have artificially synthesized organic chemicals which are not present in the natural world. These are used for a multiplicity of purposes, some of which are related to health and increasing the quality of life. At the end of 2006, the publication Chemical Abstracts listed more than 30 million different chemical compounds, and more than 12 million substances are available for purchase. Of these, the chlorinated organic compounds demand particular attention. Chlorinated chemicals have been produced and consumed on a large scale over the last 50 years, and groundwater contamination by such chemicals has been reported many times. In particular, in connection with the growth of the precision machine industry and the information technology industry, the usage of trichloroethylene and tetrachloroethylene has greatly increased. These organic chlorinated compounds have such characteristics as high specific gravity and high volatility while being uninflammable, having superior degreasing properties, low surface tension and low viscosity. These characteristics are appropriate for solvents, degreasing and dry cleaning of instruments.

An incident that occurred in Silicon Valley, California, in the mid-1980s, was widely reported by the press and was a serious issue. It concerned groundwater contamination by organic chlorinated chemicals emitted from semiconductor plants. Trichloroethylene and tetrachloroethylene are used in non-diluted concentrations, and if any contaminated waste is allowed to seep into the ground through inappropriate procedures, it rapidly penetrates the surface. The compounds may reside in material above the groundwater layer, or between upper and lower aquifers. Expansion of the contamination area is usually correlated with downward movement of the materials by the penetration of rain water, but horizontal movement also occurs through both diffusion and lateral groundwater flow. If a huge amount of non-diluted materials penetrates underground, contamination levels increase and may persist for a very long period. In addition to trichloroethylene and tetrachloroethylene, other chlorinated organic chemicals such as dichloromethane, chloroform, carbon tetrachloride, dichloroethane, trichloroethane and dichloroethylene, as well as the non-chlorinated hydrocarbons, benzene, toluene and xylene, can also   contaminate groundwater and present human health hazards.

Volatile organic compounds in groundwater were surveyed from 1985 to 1995 in USA. As part of the national water-quality assessment program of the U.S.Geological Survey, an assessment of 60 volatile organic compounds (VOCs) in untreated, ambient groundwater of the conterminous U.S was conducted based on samples collected from 2948 wells between 1985 and 1995. The samples represent urban and rural areas, and drinking water and non- drinking water wells. VOC concentrations generally were low: 56% of the concentrations were less than 1 μ g L^{-1} in urban areas, 47% of the sampled wells had at least one VOC, and 29% had two or more VOCs; furthermore, U.S. Environmental Protection Agency drinking water criteria were exceeded in 6.4% of all sampled wells and in 2.5% of the sampled drinking water wells. In rural areas, 14% of the sampled wells had at least one VOC; furthermore, drinking water criteria were exceeded in 1.5% of sampled wells and in 1.3% of the sampled drinking water wells. Solvent compounds and the fuel oxygenate methyl tert-butiyl ether (MTBE) were among the most frequently detected VOCs in urban and rural areas. The detection concentration range and frequency of 46 VOC compounds from wells in rural and urban areas are shown in Figure 1, a) and b).

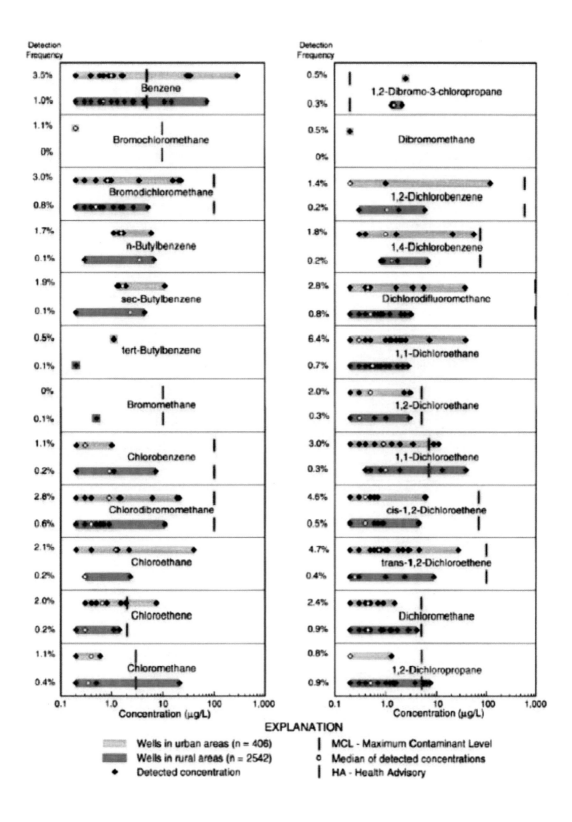

EXPLANATION

Wells in urban areas (n = 406) | MCL - Maximum Contaminant Level

Wells in rural areas (n = 2542)   o Median of detected concentrations

♦ Detected concentration | HA - Health Advisory

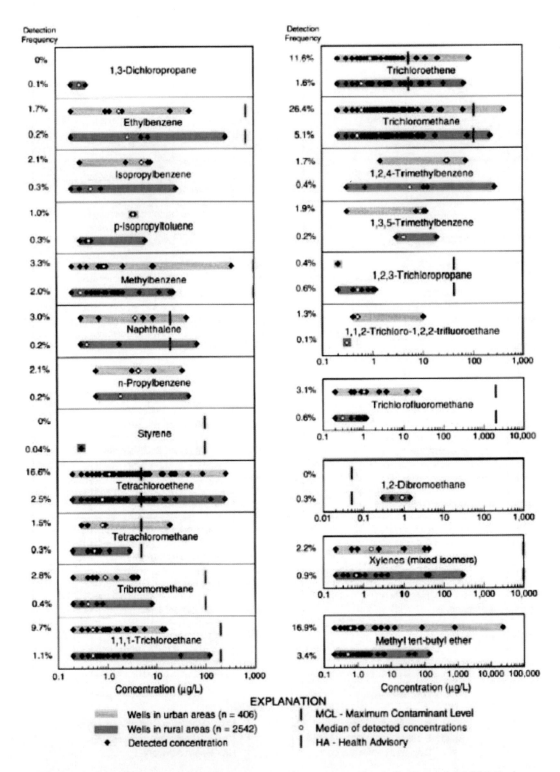

Figure 1.a) and b) .Detection frequency and concentration of selected 24 VOCs in sampled wells in urban and rural area. (Squillace P.J, E&T 1999)

Microorganisms in the soil often decompose volatile organic compounds in the groundwater. For example, the degradation pathway in the groundwater of PCE is shown in Figure 2.

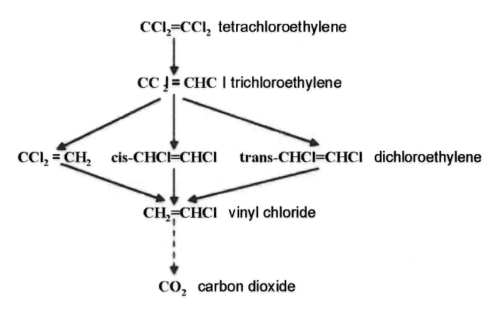

Figure2 Degradation pathway of tetrachloroethylene

By dechlorinization by microorganisms in soil, PCE decomposes to TCE. This dechlorination occurs again and TCE changes to cis- and trans-dichloroethylene. It decomposes again by dechlorination, and finally becomes carbon dioxide.

5. Agricultural Chemicals

Particularly in recent decades, the human population has markedly increased in many areas of the world. To meet the food demands of the increased population, various agricultural chemicals have been developed in order to greatly increase the harvest volumes of field crops. Agricultural chemicals can be divided into groups depending on their method of use, such as herbicides for weed control, insecticides for reducing insect damage to field crops, and fungicides to control harmful fungi. When chemicals are used, a proportion of the material is not fully decomposed and this can infiltrate into the ground in rain, leading to groundwater contamination. When discussing issues of groundwater contamination by agricultural chemicals, the possibility of permeation of the chemicals into rivers, lakes and swamps must be recognized. For example, the result of investigating pesticides in groundwater by the national water-quality assessment program (NAWQA) of U.S.A. from 1992-2001 is shown in Table 3.

Pesticides were measured in water wells such as in urban areas, undeveloped areas, for agricultural use, and groundwater in unconfined acquifers.

Seventy-six pesticides including insecticides, fungicides and herbicides, and seven decomposed substances were measured. The 47 pesticides in the 900 to 1440 water wells used for agriculture were measured using gas chromatography-mass spectrometry (GC-MS). As a result, 43 pesticides were detected. The most frequently detected pesticide was deethylatrazine, which is a degradation product of atrazine. This pesticide was detected from 42% of 1438 wells at a maximum concentration of 2.6 μ g L^{-1}. A triazine-type herbicide, atrazine, was also detected in 41.9% with a maximum concentration of 4.78μ g L^{-1}. The pesticides detected at more than 10% from well water were simazine(detected at 18.4%, 1.38μ g L^{-1}max.), metolachlor(detected at 17.0%, 32.8μ g L^{-1}max.) and prometon(detected at 9,98%, 40μ g L^{-1}max.). Hydrophilic pesticides of 36 substances, analyzed by HPLC, were measured. As a result, the detection rate was mostly low, and 13 pesticides were not detected. Pesticides detected at more than 2% in wells were only three compounds, bentazon, diuron and norflurazon, and the range of concentration was 0.01-43μ g L^{-1}.

	Agricultural L.		Aquifer survey		Undeveloped L		Urban land	
	(%)	Max.	(%)	Max.	(%)	Max.	(%)	Max.
Alachlor	2.70	0.946						
Atrazine	41.9	4.78	18.6	2.8	13.4	0.018	32.0	4.2
Carbaryl					4.48	0.02	1.56	0.031
Carbofuran	1.59	1.3						
Cyanazine	1.25	0.16						
P,p'-DDE	3.26	0.008	2.65	0.006	7.46	0.002	3.96	0.005
Dacthal	1.18	10						
Deethylatrazine	42.0	2.6	20.6	2.0	14.9	0.034	31.5	0.56
Diazinon			1.40	19			1.92	0.023
Dieldrin							5.10	5.6
2,6-diethylaniline	1.25	0.049						
EPTC	1.11	0.45						
Metolachlor	17.0	32.8	5.04	2.62	1.49	0.005	8.98	2.09
Metribuzin	3.19	0.763					1.32	0.49
Molinate					1.49	0.003		
Prometon	9,98	40	5.52	1.3	1.49	0.008	22.3	10
Simazine	18.4	1.38	6.18	0.315	5.97	0.13	18.8	1.1
Tebuthiuron	2.02	1.9	2.56	17.3			6.50	2.12
Terbacil							1.20	0.093
Trifluralin					1.49	0.004		
Aldicarb sulfone					2.13	0.007		
Bentazon	3.94	11.46	2.30	6.4			2.58	0.76
Bromacil	2.53	21.9					2.75	3.0
Dinoseb					2.13	0.02		
Diuron	3.84	5.53	2.10	1.92			3.24	2.0
Norflurazon	2.12	29.2						

Table showed more than 1% of detection frequency

(%): detection frequency per cent

Max: detected maximum concentration μg L^{-1}

Table 3. Pesticides in groundwater from various land use wells, 1991-2001

By measuring water in unconfined acquifers, 38 pesticides were detected. A low detection rate was found compared with wells for agricultural use. Pesticides detected at over 10% were 20.6% for deethylatrazine and 18.6% for atrazine. These concentrations were also low, and it was detected within 0.004-19μg L^{-1}. In hydrophilic pesticides, only two compounds, bentazon and diuron, were detected over 2%. Within the measured 47 pesticides, well water in undeveloped areas showed only nine pesticides, atrazine, carbaryl, p,p'-DDE, deethylatrazine, metolachlor, molinate, prometon, simazine and trifluralin within 1.49-13.4%. In hydrophilic pesticide, only two materials, aldicarb sulfoxide and dinoseb, were detected. Pesticides tended to be detected from groundwater near farmland in which the use frequency was clearly high from this result.

6. Pharmaceutical and personal care products (PPCPs)

In many countries such as Europe, the USA, Canada, and Japan, pharmaceuticals used for humans and animals are detected in sewage and environmental water. Pharmaceuticals are mainly detected in effluent from sewage-treatment plants, and these effluents may contaminate the surface and groundwater. As a case study, an investigation of pharmaceuticals in well water, close to a sewage-treatment plant and river, is shown in Figure 3

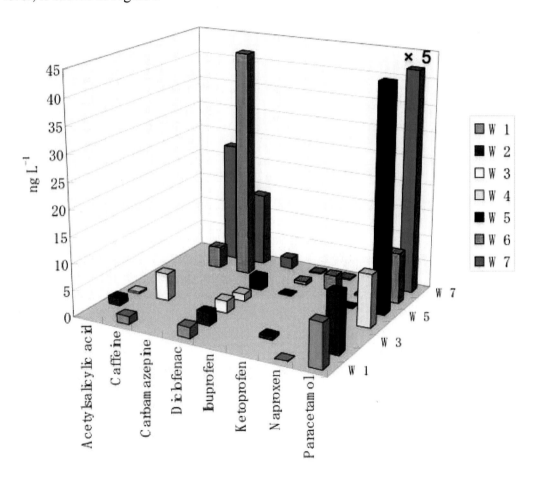

Figure3: Concentration of pharmaceuticals in the well water

As a result, eight pharmaceuticals, acetylsalicylic acid, diclofenac ibuprofen, ketoprofen, naproxen, paracetamol, an antiinflammatory drug, carbamazepine, a psychiatric drug and caffeine were detected in seven wells which supply drinking water. River and sewage effluent around the well showed higher concentrations of pharmaceuticals than well water.

7. Environmental pollution by the final disposal site leachate.

Leachate at the final disposal site may contaminate soil, groundwater and surface stream water. At the disposal site, the leachate of controlled landfill sites often shows toxic substances more than regular treatment facilities. As a result of measuring the chemical substances in the leachate at a controlled landfill site, organic substance quality of about 190 was detected, and the representative material is shown in Table 4.

Compound	Max. $(\mu g L^{-1})$	Compound	Max. (μg^{-1})
Aniline	3.38	Acenaphthene	0.005
o-chloroaniline	1.59	Acetophenone	1.60
Xylenol	1.36	Tris(butoxyethyl)phosphate	2.32
Toluene	0.18	Tris(1,3-dichloro-2-propyl)phosphate	1.89
1,4-dioxane	109	Tris(2-chloroethyl)phosphate	0.91
Xylene	0.15	Tris(2-chloropropyl)phosphate	10.9
t-butylphenol	0.45	Triethyl phosphate	0.92
Phenol	82.9	Diethyl phthalate	2.8 0
Cresol	1.99(o), 45.8(m,p)	Dibutyl phthalate	1.10
Bisphenol A	12.3	Di-2-ethylhexyl phthalate	2.51
m-toluidine	2.57	Caffeine	9.28
Methylnaphthalene	0.20	Acetic acid	46300
Benzothiazole	0.28	Isovaleric acid	3500
Dimethylnaphthalene	0.16	Isobutylic acid	3400
Naphthalene	0.035	Propyonic acid	2600
Quinoline	0.022	Simazine	1.11

Table 4. Compounds detected by the final disposal site leachate

VOC, PAH, phenols, amines, phthalates and others were detected in the leachate.

In particular, organic acid, which is produced during the process in which organic substance is resolved (i.e, acetic acid, isovaleric acid, isobutyric acid, propionic acid) was detected. The concentration showed a value as high as 2.6-46mg L^{-1}. If the management of disposal sites is not sufficient, soil and groundwater near the site may be contaminated with hazardous substances transuding from the bottom of the disposal site,

8. Environmental standard item and detection status of the groundwater in Japan.

Environmental standards have been determined for 26 toxic substances affecting health in the groundwater by the Water Pollution Control Law in Japan. The law provides that the governor of a prefecture shall monitor the quality of groundwater in accordance with the groundwater quality monitoring program prepared annually by each prefecture. The survey result carried out in 2004 is shown in Table 5.

Well water was investigated in 993 ~ 4248 locations. The environmental standards were exceeded in 387 wells, and the total excess rate was 7.8%. Materials with highest the excess rate were nitrate nitrogen and nitrite nitrogen, exceeded in 235 locations, and improved by 5.5%. There are many cases in which the pollution cause, domestic drainage or livestock waste of nitrogen fertilizer, is different to groundwater pollution by nitrate nitrogen and nitrite nitrogen. Next, arsenic showed a 2.0% excess rate in 74 locations, and in metals, the excess rate of lead was also high. In volatile organic compounds, there were many wells in which trichloroethylene and tetrachloroethylene levels were exceeded, and in addition, there was some excess of fluorine and boron. The pollution cause of fluorine and boron may be elution from the geology of natural derivation and industrial liquid wastes.

Substances	Number of wells surveyed	Number of wells exceeding EQSs	Rate of exceeding EQSs (%)	Environmental quality standard (EQS)
Cadmium	3.247	0	0	0.01 mg/L or less
Total cyanide	2.723	0	0	Not detected
Lead	3.566	14	0.4	0.01 mg/L or less
Chromium (VI)	3.420	0	0	0.05 mg/L or less
Arsenic	3.666	74	2.0	0.01 mg/L or less
Total mercury	3.235	5	0.2	0.0005 mg/L or less
Alkyl mercury	993	0	0	Not detected
PCBs	1.899	0	0	Not detected
Dichloromethane	3.535	0	0	0.02 mg/L or less
Carbon tetrachloride	3.661	4	0.1	0.002 mg/L or less
1,2-dichloroethane	3.267	0	0	0.004 mg/L or less
1,1-dichloroethylene	3.744	2	0.1	0.02 mg/L or less
Cis-1,2-dichloroethylene	3.743	5	0.1	0.04 mg/L or less
1,1,1-trichloroethane	3.990	0	0	1 mg/L or less
1,1,2-trichloroethane	3.259	1	0.0	0.006 mg/L or less
Trichloroethylene	4.234	18	0.4	0.03 mg/L or less
Tetrachloroethylene	4.248	22	0.5	0.01 mg/L or less
1,3-dichloropropene	3.043	0	0	0.002 mg/L or less
Thiuram	2.472	0	0	0.006 mg/L or less
Simazine	2.628	0	0	0.003 mg/L or less
Thiobencarb	2.539	0	0	0.02 mg/L or less
Benzene	3.524	0	0	0.01 mg/L or less
Selenium	2.698	1	0.0	0.01 mg/L or less
Nitrate-N/Nitrite-N	4.260	235	5.5	10 mg/L or less
Fluorine	3.542	19	0.5	0.8 mg/L or less
Boron	3.499	8	0.2	1 mg/L or less
Total (Actual number of wells)	4.955	387	7.8	

Table 5: Monitoring results of groundwater quality in 2004

Glossary

BOD:	Biochemical Oxygen Demand
EPTC:	S-Ethyl-N,N-Dipropylthiocarbamate
NH₄-N:	Ammonium Nitrogen
NO₂-N:	Nitrite Nitrogen
NO₃-N:	Nitrate Nitrogen
Org-N:	Organic Nitrogen
GC-MS:	Gas Chromatography-Mass Spectrometry
HPLC:	High Performance Liquid Chromatography
NAWQA:	National Water-Quality Assessment Program
PAH:	Polycyclic Aromatic Hydrocarbons
PCE:	Perchloroethylene (Tetrachloroethylene)
p,p'-DDE:	1,1-Dichloro-2,2-Bis(P-Chlorophenyl) Ethane
PPCPs:	Pharmaceutical and Personal Care Products
TCE:	Trichloroethylene
VOCs:	Volatile Organic Compounds
WHO:	World Health Organization

Bibliography

Kolpin D.W. and Martin J.D,(2003) Pesticides in groundwater: Summary statistics; Preliminary results from cycle 1 of the National Water Quality Assessment Program(NAWQA), 1992-2001, March 25. [This report describes the results of about 10 years monitoring of pesticides of groundwater in USA]

Ministry of the Environment: Government of Japan, Press release: Monitoring results of groundwater quality in 2004 [This report describe the result of monitoring of 26 substances which water quality standards in Japanese groundwater]

Rabiet M., Togola A., Brissaud F., Seidel J-L., Budzinski H. and Poulichet F.E,(2006) Consequences of treated water recycling as regards pharmaceuticals and drugs in surface and ground waters of a medium-sized Mediterranean catchment, Envion. Sci. Technol, 40, 5282-5288. [This paper shows 16 compounds of pharmaceuticals in the Mediterranean groundwater]

Squillace P.J, Moran M.J., Lapham W.W., Price C.V., Clawges R.M. and Zogorski J.S., (1999) Volatile organic compounds in untreated ambient groundwater of the United States, 1985-1995, Environ. Sci. Technol. 33, 4176-4187. [This paper describes the result of 10 years monitoring in groundwater of USA]

WHO, United Nations synthesis report on arsenic in drinking water, 2001,

http://www.who.int./water_sanitation_health/dwq/arenic3/en/. [This report shows arsenic in drinking water including source and behavior in natural waters, environmental health and human exposure assessment, drinking water guidelines and standards, safe water technology and others]

Biographical Sketch

Yoshiteru Tsuchiya is a Lecturer in the Faculty of Engineering of Kogakuin University, where he has been in his present post since 2000. He obtained a Bachelor Degree in Meiji Pharmaceutical College in 1964. He worked for Department of Environmental Health in Tokyo Metropolitan Research Laboratory of Public Health until 2000. In the meantime, he obtained a Ph.D in Pharmaceutical Sciences from Tokyo University. From 2002 to 2003, he worked for the Yokohama National University Cooperative Research and Development Center as Visiting Professor. He has written and edited books on risk assessment and management of waters. He has been the author or co-author of approximately 70 research articles. He is member of Japan Society on Water Environment, Pharmaceutical Society of Japan, Japan Society for Environmental Chemistry, Japan Society of Endocrine Disrupters Research, and International Water Association.(IWA)

SURFACE WATER DEGRADATION BY HUMAN ACTIVITIES

Yun. S. Kim
Water Analysis and Research Center, K-water, Daejeon, Korea

Keywords: environmental health, aquatic environment, endocrine disrupting chemicals (EDCs), persistent organic pollutants (POPs), organochlorine pesticides, perfluorinate chemicals (PFCs), stable isotope ratio of carbon and nitrogen, heavy metals.

Contents

1. Introduction
2. Causative materials for contamination of surface water
3. Heavy metals
4. Pollution for the phthalate esters caused from human activities
5. Endocrine disrupting chemicals
6. Persistent organic pollutants (POPs)
7. Evaluation of aquatic environment using stable isotope ratio of nitrogen and carbon in sediments and aquatic organisms
Glossary
Bibliography
Biographical Sketch

Summary

The concern of the problems caused by organic compounds in aquatic environments has been increased in the world. Surface water, which is represented by rivers, lakes and marshes, is a source of drinking water. Rain and snow that fall to the ground supply surface water. When artificially synthesized chemical substances are released into the environment, they may become part of surface runoff. Particularly, heavy metals caused by anthropogenic activities may contribute to increasing residual levels in natural waters. Furthermore, endocrine disrupting chemicals, persistent organic pollutants including dioxins and organochlorine pesticides have been extremely produced and widely used around our daily lives, since these compounds synthesized within the past 100 years. These compounds decompose slowly in the environment, and are prone to bioaccumulation in food chains. The occurrence of these chemicals in aquatic environment samples suggests harmful effects on ecosystems and human health. As clean water is essential to human health, some new analytical and treatment techniques in order to improve water quality from organic compounds are required.

1. Introduction

Surface water is the main source of drinking water. When surface waters become contaminated by pathogenic microorganisms they can cause outbreaks of infectious diseases. Recently, this point of view for the environmental health issue has been increasingly emphasized in the world. For the benefit in our daily lives, organic compounds have been extremely synthesized and produced in the past for only about a century.

Some heavy metals and organic pollutant compounds synthesized are transported for long distances from the emission sources through the atmosphere and water and residue to long terms in the nature environment. These chemicals proved to have adverse effects on ecosystems and human health. Nowadays, concern about the human health related to the environmental problems caused by the organic chemical substances in environmental samples such as water bodies, sediments and atmosphere has been continuously increasing in the world.

In here, as the emergent chemicals, endocrine disrupting chemicals, persistent organic pollutants (POPs) such as dioxins, dioxin like PCBs and organochlorine pesticides is widely produced and used around our society, since these compounds were synthesized within the past 100 years. POPs are frequently found in the Arctic despite never having been used there. It means that they are significantly persistent in the environment and widespread global distribution. Most these compounds have a hydrophobic characterization. Then, they are more easily transported with suspended substances in the water phase. The suspended substances with the absorbed organic compounds in water flow to rivers and accumulate to the sediment phase. Then, sediment cores have been used to determine modern and historical inputs of contaminants to aquatic environment.

Many industrial products are distributed in the form of unpurified compounds to our life environment and finally settled to the natural environment. In several national projects, some known estrogens such as phthalate esters (PAEs), nonylphenol (NP) and bisphenol A have been reported and investigated about their release and fate in aquatic environment. Regarding estrogenic activity as endocrine disruption, isomer-specific analysis as a new technique should be attempted because the different activity occurred depending on each isomer and distribution of their isomers in each environmental phase is different.

In the year 2001, the treaty of POPs was estimated to regulate and restrict using and producing them in the global scale at the Stockholm convention. POPs are endocrine disrupters and persistent in the environment, having long half-lives in water, air, soil, sediments or biota. Because of their lipophilic nature, they tend to accumulate in higher trophic animals through food chains.

In terms of assessing the levels of contamination, the recovery on chemical analysis of POPs in environmental samples is very important because they residue especially low level as part per billion or trillion in aquatic environment. In order to increase precision and sensitivity for chemical analysis of POPs, high-resolution gas chromatography coupled with a high-resolution mass spectrometric detector (HRGC-HRMS) is highly required and used throughout the world. To understand the emission sources of these compounds, the concentrations and the specific compositional characteristics of POPs in the environmental samples can be employed.

Moreover, nitrogen and carbon stable isotope ratios ($\delta^{15}N$ and $\delta^{13}C$) have been widely used to assess the dynamics of organic matter in biogeochemistry and ecological study. From the isotopic viewpoint, $\delta^{13}C$ and $\delta^{15}N$ might be varied depending on environmental and their own conditions. Therefore, the variation of stable isotope ratios in the environmental samples might provide basic information to understand and predict their changing.

Here, organic substances that are emerging issue are mainly described, and also organic substances in the sediments, which causes surface, water the contribution or effect is described.

2. Causative materials for contamination of surface water

Waste water from industries such as iron and steel, electroplating, chemical, paper and pulp, textiles, food, pharmaceuticals and rubber, often drains into surface waters only after removal of materials of commercial value, e.g. metals and certain organic chemicals. Domes-tic sewage is discharged to rivers after removal of organic matter and some chemicals at sew-age treatment works. Levels of nitrates and phosphates in the effluents can still be very high, however, and these can have significant adverse effects on down-stream aquatic ecosystems. Some components of domestic sewage are largely untreated, e.g. domestic synthetic detergents. These cause contamination and eutrophication of the receiving waters.

For example, when water is stored in lakes, marshes or reservoirs, it is likely to support increase of algae, some of which may produce undesirable by-products, giving unpleas-ant odor such as geosmin and 2-methylisobornenol or producing harmful effects on human health such as microcystins. Agricultural chemicals are directly and widely distrib-uted into the environment. If they are physico-chemically stable, they become causative materials for ecological changes in rivers, lakes and marshes.

Recently, the following are noticed as emergent chemicals: Phthalate, endocrine disrupt-ing chemicals (EDCs), persistent organic pollutants (POPs) and pharmaceuticals for humans and animal use.

3. Heavy metals

The pollution of the surface water by the heavy metals may become that industrial wastes such as metal plating, manufactures of pigments and paints, fungicides, the ce-ramic and glass industry and others are caused. From the metals in the surface water, there are Cd and Hg as a causing the health hazard to human. In Japan, chronic toxicity of cadmium named Itai-itai (Ouch-Ouch) disease, an osteomalacia with various grades of osteoporosis accompanied by severe renal tubular disease, and low-molecular-weight proteinuria have been reported among people living in contaminated areas around the Zintsu river basin. Humans can be exposed to cadmium via drinking water and foods from contaminated rivers polluted with the discharge of ore waste. A field survey found that the discharge of wastewater in a mine contained 4 mg L^{-1}, and river water 10 µg L^{-1} of cadmium. In river sediment, 0.16-0.18 ppm of cadmium was detected upstream of the mine, and 4.1-23.8 ppm downstream. The daily intake of cadmium in the most heavily contaminated areas amounted to 600-2000 µg day^{-1}; in other less heavily contaminated areas, daily intakes of 100-390 µg day^{-1} have been found.

Mercury in wastewater produced by chemical plants was metabolized to the organic form (methyl mercury) which is by-products of acetaldehyde synthesis. Methyl mercury caused contamination of an enclosed marine basin where they were biologically accu-mulated in fish and shellfish. Consumption of these contaminated food resources caused serious Hunter-Lassel syndrome, particularly in the local fishing community. This dis-ease became known as "Minamata Disease" after the name of the bay. Recently, mer-cury pollution in water has been reported around gold mining area such as the Amazon River basin in Brazil, Lake Victoria in Tanzania, Mindanao Island in the Philippines, Kalimantan Island in Indonesia, and Vietnam. The presence of mercury in water around gold mines is due to the use of mercury for producing Au-Hg amalgam. Gold particles such as gold dust collected from surface soil and river sediment in tropi-cal rain forests produce an amalgam when metal mercury 2-7 kg for every 1 kg of gold is added. Re-covery the mercury in the amalgam by vaporization using a burner gives pure gold. The Vaporized mercury, however, pollutes the air and water. The results of a field survey found high concentrations of mercury in mud and sediments from polluted regions, and in fish living in polluted water.

4. Pollution for the phthalate esters caused from human activities

Recently our lifestyle has been drastically changed to pursue common interests. Due to the changes, we are surrounded by many chemical products, such as the many kinds of plastics. The phthalate esters (PAEs) are representative compounds which are added into plastic products as a plasticizer. These compounds could be easily released from the sur-face of polymeric materials because the PAEs are not bonded chemically to the surface structure. The annual products` amount of PAEs has been increasing steadily. The used amount of diethylhexyl phthalate (DEHP) is 0.4-0.5 million tons per year in Europe. In Japan, approximately 0.32 million tons of DEHP were produced in 1996 and most of them used in the domestic environment. The cumulative products amounts of PAEs for the past 40 years, 1955 to 1996 in Japan reached ca. 12 million tons.

Estrogenic PAEs, dibutyl phthalate (DBP), diisobutyl phthalate (DIBP) and diisononyl phthalate (DINP) in aquatic environmental samples were investigated by the Ministry of the Environment in Japan between the year 1974 and 1996 in Japan. The concentration of these compounds including benzyl butyl phthalate (BBP), diethyl phthalate (DEP) and DEHP in aquatic environment samples is shown in Table 1.

WATER QUALITY AND STANDARDS – Vol. II – Surface Water Degradation by Human Activities – Fan, S. Tian

PAEs	Samples	1974 in Japan[*1]			1996 in Japan[*2]			1978 in Japan[*3]		
		Detection % (n)	Range	Limit of Detection	Detection % (n)	Range	Limit of Detection	(n) Ave.	Range	Limit of Detection
BBP	-	(0)	-	-	(0)	-	-	(0) -	-	-
DBP	River water	55% (375)	nd ~ 36	0.1 µg/l	17% (30)	0.21~1.4	0.2 µg/l	(41)[*4] 66.4	nd~471	0.1 ng/l
	Sediment	44% (370)	nd ~ 2.3	0.05~0.1µg/D.g	23% (30)	0.2~0.58	0.14 µg/D.g	(41) 10.7	0.1~52.1	0.1 µg/g
	Fish	37% (332)	nd ~ 2.0	0.1 µg/W.g	30% (30)	0.05~0.30	0.04 µg/W.g	(41) nd	nd	0.1 µg/g
	Rain	62% (111)	nd ~ 52	0.1 µg/l	- (0)	-	-	(0) -	-	-
	Air	- (0)	-	-	86% (15)	10~140	10 ng/m³	(13) 1.68	0.08~2.3	0.3 ng/m³
DIBP	River water	13% (375)	nd ~ 12	0.1 µg/l	0% (33)	nd	0.2 µg/l	(0) -	-	-
	Sediment	21% (350)	nd ~ 3.7	0.05~0.1µg/D.g	0% (33)	nd	0.026 µg/D.g	(0) -	-	-
	Fish	7% (312)	nd ~ 0.5	0.1 µg/W.g	(0)	-	-	(0) -	-	-
	Rain	7% (111)	nd ~ 34	0.1 µg/l	(0)	-	-	(0) -	-	-
	Air	-(0)	-	-	6% (18)	3.3	10 ng/m³	(0) -	-	-
DEP	-	-(0)	-	-	(0)	-		(0) -	-	-
DINP	River water	-(0)	-	-	0% (33)	nd	4 µg/l	(0) -	-	-
	Sediment	-(0)	-	-	0% (33)	nd	3.5	(0) -	-	-

© Encyclopedia of Life Support Systems (EOLSS)

							µg/D.g			
	Fish	-(0)	-	-	-(0)	-	-	(0)	-	-
	Rain	-(0)	-	-	-(0)	-	-	-	-	-
	Air	-(0)	-	-	0% (18)	-	72 ng/m³	(0)	-	-
DEHP	River water	40% (375)	nd ~ 15	0.1µg/l	12% (33)	4.3~6.8	3.9 µg/l	(41)*4 70.5	0.1~316	0.2 ng/l
	Sediment	61% (370)	nd ~ 17	0.05~0.1µg/D.g	46% (33)	0.18~22	0.15 µg/D.g	(34) 64.6	3.4~14.2	0.2 µg/g
	Fish	28% (332)	nd ~ 19	0.1 µg/W.g	33% (27)	0.15~0.96	0.06 µg/W.g	(20) 4.5	1~135	0.2 µg/g
	Rain	62% (111)	nd ~ 6.3	0.1 µg/l	-(0)	-	-	(0)	-	-
	Air	- (0)	-	-	61% (18)	8~323	6 ng/m³	(13) 1.36	1.4~4.1	0.4 ng/m³

*1,2 Ministry of the Environment Japan, *3 Giam, C.S. et al. (1978), *4 Surface seawater

Table 1 The concentration of phthalate esters in aquatic environment samples

In Japan, the value of DBP in water bodies was totally higher than that of DEHP in 1974. However, the value of DEHP in samples was higher than that of DBP in 1996. There are two reasons for this phenomenon. First, industrial products amount of DEHP in Japan has been increased dramatically but that of DBP had been decreased from the year 1971. Second, residues of DBP in natural environment degraded more rapidly than those of DEHP. Therefore, the historical distribution of maximum concentration detected and detection frequency on PAEs residues in aquatic environment is shown to decrease in Japan. However, the detection limit of PAEs in aquatic environment was 3.9 $\mu g\ L^{-1}$ in Japan. It should improve the sensitivity more by using high resolution method to suggest the criteria, 0.3-0.6 $\mu g\ L^{-1}$, for safe water. Comparatively, the detection limit in seawater taken from the Bay of Mexico in 1978 was at a significantly low level, 0.0001 $\mu g/L$ and those of USA and Germany are 0.01 $\mu g\ L^{-1}$ and 0.02 $\mu g\ L^{-1}$, respectively. Based on results of PAEs investigated in sediments, 1996 Japan, the concentrations of DBP and DEHP were higher than 1973's values although the tendency in river water was shown oppositely. Therefore, temporary decreasing residue concentration of PAEs in water samples could be interpreted as gradually increasing of PAEs concentration in the sediment's environment.

5. Endocrine disrupting chemicals

The biological activities of chemicals are usually isomer-specific; for example, 17beta-estradiol (E2) that is produced naturally in ovary and also synthesized artificially from estrone is strongly estrogenic, whereas its alpha-isomer is not. There are theoretically ca. 170 kinds of isomers of nonylphenol (NP), based on the structure of the nonyl side chain in NP. The structural features of alkylphenolic chemicals associated with estrogenic activity. Therefore, the estrogenic activity of alkylphenols was dependent on the carbon numbers of the alkyl chain in the phenolic structure. The fractions of a commercial NP mixture were separated using two-dimensional capillary gas chromatograph equipped with a preparative fraction collector (2DGC-PFC) and their fractions showed different estrogenic activity in a mammalian cell line, MVLN (MCF-7). Acute toxicities of NP for aquatic organisms range from 0.04 to 5 mg L^{-1} Therefore, development of simple and sensitive analytical methods for alkylphenol ethoxylates in biological samples was pointed out to be important. NP was defined as one of EDCs by the Japanese Ministry of Environment and further study would be more required.

NP is widely used for plastic additive, antioxidant, and so on. However, the major uses of nonylphenol ethoxylates (NPEs) that are produced by the addition of ethylene oxide to NP and may have 1 to 20 ethoxy units per molecule are nonionic surfactants. The NPEs are used as the surfactant in detergents, paints, emulsifying agents, pesticides herbicides and dispersing agents for agricultural and industrial applications to paper, fiber, metals and agriculture chemicals. Approximately 20,000 tons of NP and 11,300 tons of NP were annually produced in Japan and imported in South Korea, respectively. In the EU in 2004, 73,500 tons of NP was produced. Approximately, their 60% of amounts of NP in EU was used for production of NPEs, 37% for that of plastic, resin and plastic stabilizer, and the rest for that of aromatic oximes. These NP branched type has been used in our daily lives.

The NP is a degradation product of nonylphenol ethoxylates oligomers (NPEO) and exists in aquatic environment as pollutants. The NP in water samples of rivers, sewage effluents and estuaries has been widely surveyed in many countries. The NP in Canada was found, at <1 to 30 μg L^{-1} in the effluent and >100 μg g^{-1} in sludge. NP in environmental samples were collected from Lake Huron and Lake Ontario USA and Canada were detected at up to 37 μg g^{-1} in sediments and >300 μg g^{-1} in the sewage sludge. In England, the total NP in rivers and sewage effluents was detected in the range <0.2-12 μg L^{-1}. NP (n=3) between 256 and 824 mg kg^{-1} dry weight in sludge from two sewage treatment plant (STPs) in England was detected. The NP in all 149 sewage sludge samples originating from all western German states from 1987 to 1989 was found in the range from <0.5 to 1193 mg kg^{-1} dry weight, and detected at up to 83.4 mg kg^{-1} dry weight in 50% of the samples. The NP in sediment in Lake Shihwa and Masan Bay in South Korea ranged from 20.2 to 1820 μg kg^{-1} 113 to 3890 μg kg^{-1} on a dry weight basis, respectively. Residual aromatic surfactants and their degradation products in river waters in Taiwan were reported. The most abundant compounds were carboxylalkylphenol ethoxy carboxylates (19-755 μg L^{-1}), followed by nonylphenol polyethoxy carboxylates (16-292 μg L^{-1}) and NPEO (2.8-25.7 μg L^{-1}). The NP homology was significantly less abundant (1.8~10 μg L^{-1}). The NP in the river water ranged from 0.051 to 1.08 μg L^{-1} in the Sumidagawa River and Tamagawa River, Japan, while 50 to 60% of NP in the water occurred in the particulate phase. In a survey of NP and 4-tert-octylphenol in water and fish in Lake Biwa, 0.11-3.08 μg L^{-1} of NP in water was determined at high frequency (48/48). According to a report of the Japanese Ministry of Environment in 2001, the residual NP was surveyed at ND (< 0.3-0.1) to 21 μg L^{-1} in water, at ND (< 3-87) to 12,000 μg kg^{-1} in sediments and ND (< 15) to 780 μg kg^{-1} in aquatic organisms in Japan. Recently, the US EPA has prepared a guideline of total NP concentration for ambient water quality to be below 6.6 μg L^{-1} in freshwater and below 1.7 μg L^{-1} in seawater.

The analyses for NP were reported widely by HPLC, capillary-GC and GC-MS methods. Isomer-specific analysis of NP in water samples by the high resolution separation technique, GC-PFC was described a few years ago and it has generated interest with related scientists of many countries. International comparison of NP pollution in hydrospheric environment between the USA and Japan was previously reported by this author in 2001. Consequently, the GC-PFC/MS/NMR method was applied to the determination each isomer of NP in both the dissolved phase and suspended solids in water samples collected from the New York-New Jersey Harbor, USA, and also collected from Ariake Bay and some areas of Tokyo, Japan. Isomer specific quantification of individual NP isomers based on relative response factor (RRF) was developed with GC-MS in combination with steam distillation extraction. The total concentrations of NP isomers in the water samples in Ariake Bay were smaller toward offshore (11 ng L^{-1}) from river mouth (49 ng L^{-1}). Those of river water varied from ND (0.038 μg L^{-1}) to 5.4 μg L^{-1} in the Tokyo area. The significant variation coefficient of the isomer proportion of NP among three marine samples varied at 4~75% while those among 16 river samples did at 11~38%. This suggests that NP isomers in aquatic environmental samples might be independently degraded, because a composition of isomers of NP was found to vary in environmental samples and individual isomers have different activities. Therefore, an estrogen equivalent concentration of NP in the environment should be calculated. The total estrogenic activities of NP in Ariake Bay seem to have been over-estimated 2.7 to 3.0 times more than the actual to total NP estrogenic activities.

6. Persistent organic pollutants (POPs)

6.1. Organic chlorinated chemicals

The most important issue is contamination of surface water by organic chlorinated compounds which are very stable and widely distributed in the environment. Most POPs except dioxins (PCDD/Fs) and PCBs are organochlorine pesticides, such as dichlorodiphenyl trichloroethane and its metabolites (DDTs), hexachlorobenzene (HCB), heptachlor, chlordane compounds (CHLs), aldrin, dieldrin, endrin, mirex and isomers of hexachlorocyclohexane (HCHs), which are nominated as new POPs in the near future. These compounds are suspected carcinogens from the point of view for the environmental health. The chemical properties of these POPs are typically lipophilic and persistent nature. Organic chlorinated chemicals have been massively used on the agroenvironment in the world because of their strong insecticidal, herbicidal, and fungicidal effects. These chemicals leave a residue in the environment for a long time and can adversely influence ecosystems by causing contamination of rivers, lakes and marshes. To avoid this, use of these chemicals is prohibited in advanced countries, but they are still used on a large scale in developing countries, causing serious contamination. Recently, the result that the concentrations of agrochemicals POPs including PCBs in porewater were higher than those in surface water and the most predominant phase was in surficial sediment samples reported in China. It might be assumed that these hydrophobic compounds tend to residue strongly in the sediments phase, and then re-suspended from the sedimentary phase to the upper water.

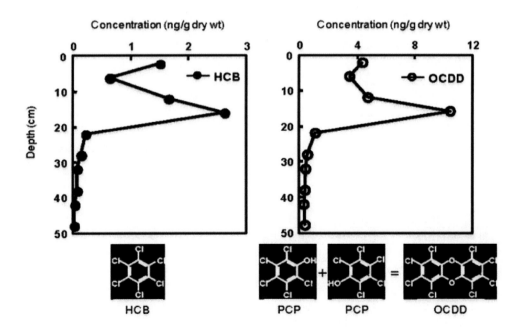

Figure 1 Vertical distributions of concentrations of HCB and OCDD in sediment core, Japan

Emission sources of HCB were considered as follows; impurities in pesticides such as PCP, pentachloronitrobenzene, dimethyl-tetrachloroterephthalate, incineration of HCB containing waste from tetrachloroethylene production and incineration of municipal solid waste. Homologue profiles of PCP samples revealed that the most abundant impurity of dioxin was octachlorinated dibenzo-*p*-dioxin (OCDD). Vertical distributions of HCB and OCDD concentration were presented as shown in Figure 1. The historical concentration profile of the HCB is shown corresponding to that of OCDD in the sediment core that it was taken from Japan. Thus, the concentration of HCB observed in the environment samples could be assumed to be attributed to the herbicide, PCP.

		Surface water (47 sites)			Sediment (63 sites)		
		CR	Ave.	DN	CR	Ave.	DN
1	PCBs	140-7800	520	47	42-690000	7500	63
2	HCB	6-210	21	47	13-22000	160	63
3	Aldrin	ND-5.7	Tr (0.6)	32	ND-500	7.5	62
4	Dieldrin	4.5-630	39	47	2-4200	56	63
5	Endrin	ND-120	4.0	45	ND-19000	10	61
6	DDTs						
6-1	*p,p'*-DDT	1-110	8	47	5.1-1700000	280	63
6-2	*p,p'*-DDE	4-410	26	47	8.4-64000	630	63
6-3	*o,p'*-DDD	1.8-130	17	47	5.2-210000	520	63
6-4	*o,p'*-DDT	ND-39	3	42	0.8-160000	47	63
6-5	*o,p'*-DDE	0.4-410	2.5	47	ND-31000	35	62
6-6	*o,p'*-DDD	0.0-51	5.2	47	0.8-32000	110	63
7	Chlordanes						
7-1	*cis*-Chlordane	6-510	53	47	3.3-44000	140	63
7-2	*trans*-Chlordane	3-200	25	47	3.4-32000	98	63
7-3	Oxychlordane	ND-19	2.6	46	ND-160	2.1	51
7-4	*cis*-Nonachlor	0.9-43	6.0	47	1.1-9900	50	63
7-5	*trans*-Nonachlor	2.6-150	20	47	2.4-24000	89	63
8	Heptachlors						
8-1	Heptachlor	ND-54	ND	25	ND-200	2.5	48
8-2	*cis*-Heptachlor epoxide	1.0-59	7.1	47	ND-140	4	49
8-3	*trans*-Heptachlor epoxide	ND	ND	0	ND	ND	0
9	Toxaphene	ND	ND	0	ND	ND	0
10	Mirex	ND-1.0	ND	14	ND-5300	1.5	48
11	HCHs						
11-1	α-HCH	16-660	90	47	3.4-7000	120	63
11-2	β-HCH	25-2300	200	47	3.9-13000	180	63
11-3	γ-HCH	8-250	48	47	1.8-6400	44	63
11-4	δ-HCH	ND-62	1.8	23	ND-6200	46	63

CR= Concentration range, Ave.= Average concentration, DN= Detection Number, Concentration units; water pg/L, sediment pg/g-dry

Table 2 The concentration of POPs in surface water and sediments in Japan (2005)

The HCHs residues are among the most widely distributed and frequently detected organochlorine contaminants in the environment. The nationwide HCHs usage in Japan was approximately 400 metric tons. This is an extremely large amount when compared with DDTs and PCBs in Japan. Interestingly, production of HCHs in Japan was highest in the early 1970s and the trend of the residues level investigated in biota was dependent on its usage. The most residual concentration of HCHs in the sediment core was significantly shown to have similar variation in these corresponding ages. Therefore, it can be interpreted that HCHs contamination might have occurred through the use of technical-grade HCHs until the early 1970s. Further, the tendency of HCHs concentration in surface water showed to be decreasing in the world. It might be assumed that manufacturers and use of the technical HCHs were banned since the early 1970s. A high percentage of α- isomer in the environment was generally reflecting the composition of technical-grade HCHs (α; 70%, β; 9%, γ; 14%, δ; 7%). The composition of β-HCH has been reported predominant in the environment because it is most stable among HCH isomers and resistant to microbial degradation.

The DDTs are under restricted use for disease vector control in the world. The technical DDTs contain 75% p,p'-DDT, 15% o,p'-DDT, 5% p,p'-DDE and less than 5% others. The ratio of (DDD+DDE)/ΣDDTs could also be used powerfully to find their sources and fate in the environment. Microbial degradation of DDT, DDD and DDE is generally slow, resulting in environmental persistence of these compounds and DDT may degrade to DDD with a half-life of a few days under certain conditions. In summary then, the composition of organochlorines and their metabolities can provide some information for a better understanding of the origin and transport of these contaminants in the environment.

The result of POPs concentration in water and sediment in Japan is shown in Table2. The water samples were mainly collected from nationwide 47 sites which are near estuary in the river, and 63 sites of sediments were also collected. The POPs substances, which already stopped the use about 30 years ago, were still detected from water and sediments, and especially, the high concentration of PCBs, DDTs and chlordanes was detected in the sediments.

6.2. Organic phosphorylated chemicals

Organic phosphate triester chemicals have ester bonds of ortho-phosphate and alcohols in their chemical structures. The typical ones being used are trimethyl phosphate, triethyl phosphate, tributyl phosphate, tris(2-chloroethyl) phosphate, triphenyl phosphate, and tris(2-ethylhexyl) phosphate, and triscresyl phosphate. Upon disposal of plastic products, such materials can be released into surface water. They are known to be toxic to fishes, and there are dangers that they could affect the health of people who eat the contaminated fish, or drink the water. Some kind of these chemicals such as parathion, methylparathion and methyldimethon cause issues of contamination of surface water due to their stability in the environment and accordingly use of them are prohibited in advanced countries. As alternatives, other chemicals such as marathion and fenitrothion, which have a low residual characteristic and low toxicity to mammals, are used instead. The guidelines for drinking water quality in South Korea was estimated 0.02 mg L^{-1} for diazinon, 0.04 mg L^{-1} for fenitrothion and 0.06 mg L^{-1} for parathion in waters.

6.3. Other agricultural chemicals

Such chemicals as carbamate-group and pyrethroid-group compounds are readily decomposed in the environment and have low residual abilities; they are therefore widely used as herbicide and insecticides. However, they are directly released into the environment through their ordinary usage, and accordingly their uses should be minimized as much as possible.

6.4. Perfluorooctane sulfonate (PFOS) and perfluorooctanoate (PFOA)

PFOS and PFOA among the perfluorinated chemicals (PFCs) have been recognized as widespread contaminants and detected globally in the environment. They generally exist in ionized form because of their strong acidity in the environment. Based on the high energy of carbon to fluorine bond, PFCs are enough to resist hydrolysis, photolysis, biodegradation and metabolism in the environment and thus result in a high degree of environmental persistence and bioaccumulation of PFCs.

Recent ecological investigations have shown that one of the ecological sources of PFOS and PFOA is discharge waters from manufacturers. PFOS is utilized in fire-fighting foams, herbicide and insecticide. PFOA is widely used in various industrial applications. The concentrations of PFOS and PFOA in surface water everywhere Japan ranged from 0.89 ng L^{-1} to 5.73 ng L^{-1} and 0.97 ng L^{-1} to 21.5 ng L^{-1}, respectively. The contaminations of such chemicals show large geographical difference in their levels. Further, the contamination maximum levels of PFOS and PFOA were significantly greater in surface water than in tap water (12 ng L^{-1} for PFOS, 40 ng L^{-1} for PFOA) in Japan.

7. Evaluation of aquatic environment using stable isotope ratio of nitrogen and carbon in sediments and aquatic organisms

Surface water flows over the ground, it gains dissolved components from decomposition of plants, as well as live phytoplankton. Then the results of historical trends of stable isotope ratios of nitrogen and carbon in sediments core taken from the environment could be proofed and found about a behavior and of pollution by human activities. In summary, using this technique is powerful to interpret their sources, movement/transport and fate in environment. As a case study, evaluation of discoloring phenomena of lavers during *in situ* cultivation using the stable isotope ratios of nitrogen and carbon carried out by this author with colleagues of Nihon University, Japan. We were interested in knowing whether δ-values in the lavers could be used to check the physiological status of lavers and to predict the discoloring phenomenon. The stable carbon, nitrogen isotope ($\delta^{13}C$, $\delta^{15}N$) and C/N ratios of lavers cultivated in Ariake Bay, Japan were analyzed. Phytoplanktons exhibit regional characteristics in their $\delta^{13}C$ and $\delta^{15}N$ values according to circumstantial conditions and growth physiology. The $\delta^{13}C$ value of dissolved inorganic carbon (DIC) may be largely affected by the chemical enhancement of CO_2 invasion from the atmosphere, which shows the large negative kinetic fractionation, compared with the equilibrium of the CO_2 exchange. The stable isotope ratio, $\delta^{13}C$ or $\delta^{15}N$ was expressed as: δ (‰) = $(R_{sample} - R_{standard}) / (R_{standard}) \times 1000$, where R denotes $^{13}C / ^{12}C$ for $\delta^{13}C$ and $^{15}N / ^{14}N$ for $\delta^{15}N$. Pee Dee Belemnite and atmospheric nitrogen were the primary standards for $\delta^{13}C$ and $\delta^{15}N$, respectively.

The $\delta^{13}C$ value was -23.1 ± 0.57‰ (mean ± 1 SD) in the normal lavers period, -23.6 ± 0.77‰ in the fertilized period and -22.0 ± 0.20‰ for the period of discoloration. Aquatic plants such as macrophytes and many phytoplanktons mainly assimilate free CO_2 in water through the C_3 cycle and the HCO_3^- which is another possible substrate for phytoplankton. For the $\delta^{13}C$ values, when the phytoplankton biomass increases, the dissolved CO_2 being carbon source of the lavers in the cultural ground is depleted. Then the lavers assimilate HCO_3^- which has a higher $\delta^{13}C$ level than the co-existent CO_2. Thus, the $\delta^{13}C$ values in the lavers increase proportionately with the levels of discoloration.

The $\delta^{15}N$ value was 13.6 ± 2.6‰ in the period of normal growth, 5.9 ± 2.2‰ in the fertilized period using $(NH_4)_2SO_4$ with $\delta^{15}N$ of -2.8‰, and 1.4 ± 0.80‰ in the period of discoloration. For the $\delta^{15}N$ value, when the phytoplankton biomass increases, the inorganic nitrogen pool in the seawater which is the source of nitrogen for lavers is depleted. In this situation, the lavers assimilate both the low $\delta^{15}N$, which is supplied as a chemical fertilizer, and the land-derived nitrogen compounds. Therefore, the $\delta^{15}N$ value in lavers is significantly lowered. The C/N ratio averaged 5.5 ± 0.30 in the normal period, 6.0 ± 0.90 in the fertilization period, and 9.8 ± 1.2 in the discoloration period. These results implied that Ariake Bay is a nitrogen-deficit environment.

The variation patterns of the $\delta^{13}C$ and $\delta^{15}N$ values in CY 2002 significantly agreed with the tendency that was observed in CY 2000 when seaweed farming was severely damaged by nutrient deficiency due to extra-ordinary phytoplankton blooming. The C/N ratio of the fronds was increased with discoloring, indicating that the physiological status of the lavers changed under limited nitrogen conditions. The abrupt negative change of $\delta^{15}N$ indicated that the lavers ingested $(NH_4)_2SO_4$ which was supplied as nutrient under dissolved inorganic nitrogen deficient conditions. As a result, the $\delta^{15}N$ value in lavers was decreased. Then, the fronds were able to keep the normal color as long as possible in the period of cultivation. The fronds, however, finally turned pale. It is implied that their nitrogen had been much more depleted under the nitrogen limitation and the faded fronds did not return to normal even after fertilization.

In addition, the possible cause for the decline of biological productivity in Ariake Bay was also investigated using $\delta^{15}N$ and $\delta^{13}C$ of a 50-cm long sediment core. The overall analytical precision was ± 0.1‰ for $\delta^{13}C$ values and less than ± 0.2‰ for $\delta^{15}N$ values. Relationship among the $\delta^{15}N$ values of the sediment, lavers (Japanese: *nori*) products and number of dam construction presented in Figure 2. The core sample was categorized in three parts from the top toward the bottom. In the first segment of the 1986-2002 period, $\delta^{15}N$ of sediment showed an increasing tendency (0.03‰ per year), which was accompanied by a decrease in laver production. In the second segment of the 1970-1985 period, $\delta^{15}N$ of sediment was relatively constant at 6.8 ± 0.1‰. In the third segment of the 1950-1970 period, $\delta^{15}N$ fluctuated significantly.

The above vertical change could be related to decadal change in the number of dam constructions that might interrupt terrestrial nitrogen input to the Ariake Bay. The nitrogen depression tendency was also indicated by temporary change in $\delta^{15}N$ of edible lavers during the *in situ* cultivation period. The $\delta^{15}N$ value and the C/N ratio are applicable to monitor physiological conditions of lavers which may help to prevent lavers from discoloring. Therefore, in order to accurately evaluate the environmental changes, a stable isotopic application was acceptable and useful.

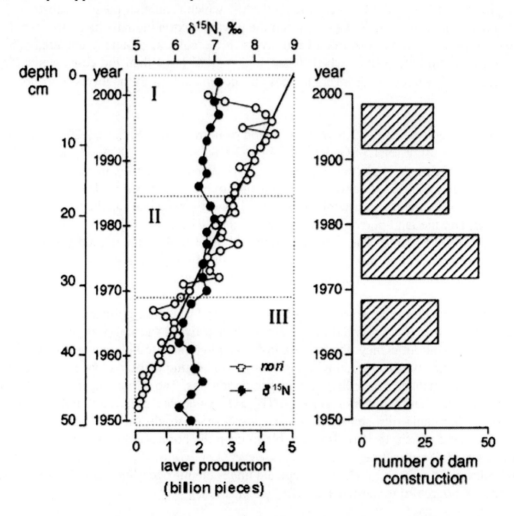

Figure 2 The $\delta^{15}N$ of dated core sediment and annual lavers (*nori*) production and number of dam construction in Ariake Bay, Japan.

Glossary

EDCs:	Endocrine Disrupting Chemicals
POPs:	Persistent Organic Pollutants
PAEs:	Phthalate Esters
GC-HRMS:	Gas Chromatography-High Resolution Mass Spectrometry
HPLC:	High Performance Liquid Chromatography
GC-PFC:	Gas Chromatograph equipped with a Preparative Fraction Collector
DEP:	Diethyl Phthalate
DBP:	Dibutyl Phthalate

BBP:	Benzyl Butyl Phthalate
DEHP:	Diethylhexyl Phthalate
DIBP:	Diisobutyl Phthalate
DINP:	Diisononyl Phthalate
NP:	4-Nonylphenol
OCDD:	Octachlorinated Dibenzo-*P*-Dioxin
PCBs:	Polychlorinated Biphenyls
HCB:	Hexachlorocyclohexane
DDT:	Dichlorodiphenyl Trichloroethane
HCHs:	Hexachlorocyclohexane
CHLs:	Chlordane Compounds
PFOS:	Perfluorooctane Sulfonate
PFOA:	Perfluorooctanoate
$\delta^{15}N$:	Nitrogen stable isotope ratio
$\delta^{13}C$:	Carbon stable isotope ratio

Bibliography

Katase T., (2004) The past and current of phthalate esters, Koudo Publication, Japan. [This book describes the review of phthalate esters during the past about 30 year's investigation in Japan]

Katase T., Kim Y.-S., Fujimoto Y. and Inoue T., (2006) Isomer-specific analysis of 4-nonylphenol with estrogenic activity and their distribution in aquatic environment in relation to endocrine disrupters, edited in Trenda in Environmental Reserch, p119-162, Nova Science Publications inc.,NY, [This book describes the structure elucidation, separation isomers, testing for estrogen activity and distribution of 4-nonylphenol in environmental waters in Japan]

Kim Y.-S., Eun H., Katase T. and Fujiwara H., (2007) Vertical distributions of persistent organic pollutants (POPs) caused from organochlorine pesticides in a sediment core taken from Ariake bay, Japan, Chemosphere, 67, 456-463. [This paper describes the historical variation of POPs in sediment core, Japan during the past about 100 years]

Saito N., Harada K., Inoue K., Sasaki K., Yoshinaga T. and Koizumi A., (2004) Perfluorooctanoate and perfluorooctane sulfonate concentrations in surface water in Japan, J. Occup. Health, 46, 49-59. [This report describe the result of monitoring of PFOS and PFOA in surface water and tap water taken from all over Japan]

Zhang Z.L., Hong H.S., Zhou J.L., Huang J. and Yu G., (2003) Fate and assessment of persistent organic pollutants in water and sediment from Minjiang River Estuary, Southeast China, Chemosphere, 52, 1423-1430. [This paper shows the distribution of selected organochlorine pesticides and PCBs in water, porewater and sediments, China]

Biographical Sketch

Yun. S. Kim is a senior researcher in the water analysis and research center of K-water, Daejeon, Korea. He obtained his Bachelor`s Degree, Master`s Degree and Ph.D (Bioresource Sciences; Environmental Analysis Chemistry) from Nihon University, Japan. He has worked with persistent organic pollutants (POPs) such as, dioxins, PCBs and organochlorine pesticides in riverine and estuarine samples and their risk assessment in East Asia at the National Institute for Agro-Environmental Sciences and the Yokohama National University, Japan, respectively. He is interested in researching structure elucidation and the development of analysis methods for emerging organic compounds in environmental samples and their risk assessments.

POLLUTION SOURCES

Yuhei Inamori
National Institute for Environmental Studies, Tsukuba, Japan

Naoshi Fujimoto
Faculty of Applied Bioscience, Tokyo University of Agriculture, Tokyo, Japan

Keywords: water pollution, eutrophication, organic matter, nitrogen, phosphorus, heavy metals, organochloride compounds, dioxins, endocrine disruptor, agricultural chemicals, domestic wastewater, industrial wastewater, acid rain, sewage works

Contents

1. Introduction
2. Pollution of organic matter, nitrogen and phosphorus
3. Pollution of heavy metals and inorganic compounds
4. Pollution of harmful organic compounds
5. Countermeasures against water pollution
6. Other forms of pollution
Glossary
Bibliography
Biographical Sketches

Summary

There are two types of pollutants, those affecting the natural environment such as organic matter, nitrogen and phosphorus, and those that affect human health. High levels of organic matter in water bodies cause depletion of dissolved oxygen and damage to other ecological conditions needed by aquatic life, offensive odors and deterioration of landscape. High levels of nitrogen and phosphorus cause eutrophication, leading to many problems such as toxic algal blooms, problems in water treatment works, deterioration of landscape and depletion of dissolved oxygen in the bottom layer. Organic matter, nitrogen and phosphorus enter water bodies mainly as domestic wastewater, industrial wastewater, treated water from sewage treatment plants, wastewater from livestock farms, and wastewater from fish farms.

There are many kinds of harmful compounds in the environment encompassing human activities. Harmful compounds that damage human health include heavy metals, dioxins, endocrine disruptors, and agricultural chemicals—particularly organochlorine compounds. These are discharged from industry, landfills, agricultural fields and incinerators. To protect human health and satisfy water quality for utilization purpose, standards have been established for environmental quality and effluent standards in many countries.

Another form of water pollution is the thermal pollution caused by conventional and atomic power plants. This has various adverse effects on aquatic life. The global environmental problems relating to the aquatic environment include acid rain. This is caused by emission of SOx and NOx from industries and automobiles.

1. Introduction

The world's water is broadly classified into seawater, lake and marsh water, river water, groundwater, soil water, atmospheric water, glaciers and ice sheets, and the water in living organisms. Around 97% of it is seawater. The water on Earth is constantly being cycled under the driving force of evaporation of water from the sea and, to a lesser extent, the land surface, followed, of course by precipitation. Water flowing on the ground as a result of rainfall flows through forests and other land, runs over the surface, becomes groundwater, and eventually flows into rivers and hence to the ocean. It flows through rivers into lakes, marshes, and reservoirs, before reaching the ocean. This is the naturally occurring background circulation of water that puts carbon, nitrogen, and phosphorus loads onto water bodies. In addition to this, human actions create an artificial element in this circulation—we obtain and purify water from sources such as lakes and marshes, reservoirs, and rivers, and we use it as domestic, industrial, or agricultural water, then discharge it into water bodies, either directly or after treatment. Most of the pollution of lakes, marshes, rivers, and the ocean is a result of the circulation of water driven by human factors.

The main pollution sources that contaminate public water bodies, rivers, lakes, and oceans, are industrial wastewater, domestic wastewater, treated water from sewage treatment plants, wastewater from the livestock industry, and wastewater from fish farms. Others are runoff from cities, seepage from industrial waste material treatment plants, seepage from waste material that has been disposed of illegally, excrement from free-ranging animals, and runoff from forests, agricultural land, and golf courses. These latter cannot be specified as individual pollution sources. Pollutants common to many of these pollution sources are organic substances, nitrogen, and phosphorus. Other pollutants are heavy metals, organochlorine compounds, agricultural chemicals, dioxins and endocrine disruptors. Agricultural chemicals that are discharged from agricultural land and golf courses can pollute both groundwater and surface water bodies.

Pollutants in public waters can be classified as those that harm the wider environment and those that are a threat to human health. Major substances that harm the environment are organic matter, nitrogen, and phosphorus. Inflow of organic matter causes bacterial growth and reduces dissolved oxygen concentration, and inflow of nitrogen and phosphorus causes eutrophication that can lead to algal blooms. These are damaging to aesthetic and recreational values and can cause major problems in water treatment processes. As eutrophication advances, cyanobacteria (blue-green algae) propagate, and these can produce hazardous substances for animal and humans. Elevated concentrations of nitrogen and phosphorus not only damage aquatic ecosystem, but are also a threat to human health.

Typical pollutants that harm human health include heavy metals, toxic inorganic compounds, organochlorine compounds, agricultural chemicals, dioxins and endocrine disruptors. If people ingest these substances, they may suffer acute or chronic intoxication, according to the exposure condition and quantity ingested. Because such substances are extremely harmful, effluents from factories and business establishments are strictly regulated by effluent standards, but there are dangers of water resources being polluted by seepage of such substances from industrial waste material treatment plants or illegally dumped industrial waste. Because many new chemicals are now being used without adequate evaluation of their harmful effects on people, wastewater regulations have not kept up with actual risk. Pollution by these new chemicals is, therefore, an extremely serious problem. In addition to pollution caused by substances, thermal pollution caused by conventional thermal and atomic power plants also affects aquatic life, particularly in rivers and coastal areas. Global environmental problems include the NOx and SOx emitted from industries and automobiles, that cause the so-called acid rain. This decreases the pH of water bodies and affects aquatic life.

2. Pollution of organic matter, nitrogen and phosphorus

2.1. Adverse effects of organic matter, nitrogen and phosphorus

2.1.1. Organic matter

Waste water from domestic, industrial and business establishments contains organic matter such as carbohydrates, fat, protein and organic acid. When these wastewaters flow into water bodies without treatment, organic matter is assimilated by aerobic microorganism such as bacteria and fungi, and dissolved oxygen is consumed with their growth. If consumption of dissolved oxygen is greater than the supply of oxygen from air and photosynthesis, the concentration of dissolved oxygen decreases. Thus organic matter pollution causes a decrease of dissolved oxygen and, as the concentration approaches zero, production of hydrogen sulfide by anaerobic bacteria begins. The living conditions for aerobic aquatic life deteriorate and species that can survive without dissolved oxygen become dominant. Water bodies with low levels of dissolved oxygen and high levels of anaerobic bacteria decomposing organic matter, often have an unpleasant smell. Fish can only survive where the dissolved oxygen concentration is at least 3 mg/l. Organic matter pollution exacerbates the odor problem, giving the water an unpleasant smell and taste, which creates problems for public water supply. The growth of microorganisms is also stimulated, and water bodies with high levels of bacteria are not suitable for recreational use.

2.1.2. Nitrogen and phosphorus

Nitrogen and phosphorus are essential elements for algae, bacteria, protozoa and metazoa including fish, i.e. the whole hydrospheric ecosystem. Excessive inflow of these elements, however, eutrophicates lakes and coastal areas. Efforts to remedy the organic pollution of water bodies in Japan focused on reducing the biochemical oxygen demand (BOD) of household and industrial wastewaters. As a result, organic pollution was significantly reduced. At the same time, however, there was higher input of nitrogen and phosphate from sources independent from wastewater treatment processes. This raised their concentrations in public water bodies, triggering frequent outbreaks of algal blooms.

Algal blooms became commonplace in the summer in eutrophic lakes and reservoirs in Japan, such as Lakes Kasumigaura, Teganuma, Inbanuma, and Suwa. The major constituent of algal blooms is cyanobacteria, represented by *Microcystis*, *Anabaena*, and *Oscillatoria*. Under appropriate conditions, these algae propagate rapidly, absorbing nitrogen from nitrate and ammonium, and phosphorus from phosphate. Nitrogen and phosphorus are provided by influent rivers, recycling of material in the bottom layer (associated with bacterial decomposition of detritus and excreta of zooplankton), elution from the sediment, and rainfall. The growth rate of algae greatly depends on light intensity, water temperature, and nitrogen and phosphorus concentrations. In addition to cyanobacteria, green algae and diatoms multiply in eutrophic lakes and reservoirs, where the dominant species are subject to seasonal change. Such seasonal changeovers happen because each algal species has different growth characteristics as a result of variation in nitrogen and phosphorus concentrations, illumination, and temperature. These factors determine which algae dominate under changing environmental conditions. The cyanobacteria are generally not dominant in oligotrophic environments, because their growth rate is low or null in oligotrophic water. It was reported that *Microcystis*, a typical cyanobacteria, dominates lakes when the total nitrogen and total phosphorus are over 0.5 mg/l and over 0.08 mg/l, respectively.

Due to their shape, e.g. colonies of cells in *Microcystis* and filaments in *Oscillatoria*, the cyanobacteria are not susceptible to being eaten by zooplankton, which is also a contributing factor in algal domination. Furthermore, cyanobacteria can float due to their intracellular gas vacuole. When they dominate the lake, the cyanobacteria cluster on the surface, which greatly reduces sunlight penetration, reducing insolation to other algae. This tendency to float accelerates the domination of cyanobacteria, unless wind or other phenomena increasing mixing of the lake.

In waterworks, elimination of suspended solids (SS) from raw water taken from eutrophic lakes and reservoirs requires the coagulation-sedimentation process. Coagulants, such as aluminum sulfate, have to be increased in proportion to the SS proliferation. Growth of algae inhibits coagulation by raising the pH level. Proliferation of algae therefore requires not only coagulants in quantity, but also may hinder the coagulation-sedimentation process. In addition, the cyanobacteria that form the water bloom tend to float, which means that they may flow out rather than settle.

Pre-chlorination is sometimes practiced to ensure sufficient chlorine disinfection by averting the obstacles to coagulation and filtering. In this process, hypochlorous acid (HClO) reacts with organic matters derived from algae to form trihalomethanes (chloroform, bromodichloromethane, dibromochloromethane, and bromoform) that are suspected of being carcinogens. Animal tests have demonstrated that, among them, chloroform and bromodichloromethane are carcinogenic, which suggest that they are also likely to be so for humans. The Waterworks Law in Japan stipulates the Water Quality Standards for Drinking Water for all four substances as well as their total value (total trihalomethanes) as 0.1 mg/l.

Algal blooms triggered by eutrophication clog the filters. The clogging-causing algae are not limited to cyanobacteria; blooms of diatoms, such as *Synedra* and *Melosira*, are also found to cause clogging. Algal blooms can also impart musty odors to drinking water. Some are effused by algae directly, while others are produced through the decomposition of dead algae, affected by actinomycetes and bacteria. Musty odors are caused by 2-MIB (2-methylisoborneol) and geosmin, which are found to be effused by *Phormidium tenue*, *Oscillatoria* and *Anabaena*.

Massive amounts of algae, spawned by eutrophication, die and settle out on the bottom to undergo bacterial decomposition. As a result, the bottom layer becomes anaerobic, causing iron and manganese elution from the sediment. These metals in tap water can stain laundry and affect the taste of the water. Consequently, they need to be eliminated through a water purification process. Among the water-coloring metals, the Water Quality Standard of Drinking Water under the Waterworks Law in Japan regulates manganese and aluminum. The WHO Guidelines for Drinking Water Quality sets guideline values for aluminum, copper, iron, and manganese.

Growth of phytoplankton in eutrophic lakes reduces the transparency, coloring the lakes green in the case of cyanobacteria and brown with diatoms. This significantly reduces the quality of the aquatic environment. When the algal bloom forms a surface scum, the lake is unfit for recreation activities such as swimming, water-skiing, and boating. The opacity of the water is itself a danger, but it is also important to remember the toxins released by cyanobacteria. A case was reported from England involving soldiers who became ill after canoe-training in water with a heavy bloom of *Microcystis*. Water blooms can give off offensive odors, making the area less suitable for hiking and other forms of recreation, and any nearby residents may be distressed by foul odors from decomposing algae and hydrogen sulfide from the anaerobic bottom layer, caused by organic decomposition.

Water taken from eutrophic lakes and rivers for agricultural irrigation can have an adverse effect on crop production. Nitrogen in particular significantly reduces rice yield through overgrowth, lodging, poor maturation, and frequent pest outbreaks. It is normally considered that 1 mg/l of nitrogen is harmless, whereas a level exceeding 5 mg/l causes grave problems. When mostly-toxigenic cyanobacteria bloom in ponds or lakes to which livestock have access, they may develop health disorders as a result of drinking infested water. Cases of lethal cyanobacterial intoxication of livestock, including cattle, sheep, pigs, and fowl, have been reported from Australia and USA. Furthermore, an accident was reported from Brazil in which 54 people died after drinking tap water containing the toxic substance microcystin.

2.2. Pollution source of organic matter, nitrogen and phosphorus

2.2.1. Domestic wastewater

From domestic sources, pollutants are mainly generated from excretion, cooking, washing and bathing. Major pollutants are organic matter, nitrogen and phosphorus. Beside excrement, they come from cooking oil, detergent, body soap and food. In Japan, it was calculated that one person per day generates approximately 200 L of wastewater, containing 40g, 10g, and 1 g for BOD, nitrogen and phosphorus respectively (see Figure 1). Averaged water quality of domestic wastewater is BOD 200 mg/l, T-N 50 mg/l and T-P 5 mg/l.

Significant pollution loads are generated by kitchens, from leftovers, waste broth and boiling water, sauces and seasonings, oil, and detergent from washing of dishes. Table 1 shows water quality of selected sauces, alcoholic drinks, wastewater from washing rice and oil. It is obvious that BOD, N and P are very high in seasonings, alcoholic drinks and oil. For reducing wastewater load from kitchens, waste of these things should clearly be minimized. From bathrooms, pollution loads are caused by body soap, shampoo, rinse, perspiration and grime.

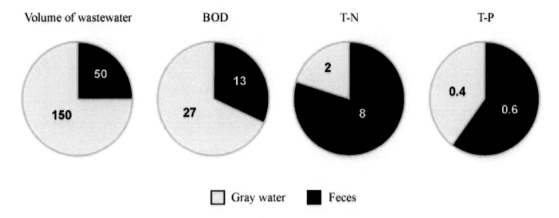

Figure 1. Pollution sources from domestic water use

	BOD	T-N	T-P
Soy sauce	220,000	25,000	3,900
Beer	89,000	340	132
Fruit juice	110,000	420	110
Waste water from washing rice	2,400	29	7.8
Water from boiling spaghetti	5,400	55	17
Cooking oil	1,670,000	1,400	30

Table 1. Equivalent water quality of selected sauces, alcoholic drinks, wastewater from washing rice, and oil

Pollution loads from human feces are important because the proportion of feces in the pollution load is very high. As feces, one person discharges an average BOD of 13 g, nitrogen 8 g, and phosphorus 0.6 g per day (Figure 1).

If domestic wastewater is treated effectively by a sewage works or small scale domestic wastewater treatment facility, the pollution load from domestic water use can be greatly reduced. However, if domestic wastewater is not treated and flows directly into water bodies, pollution, including decrease of dissolved oxygen, bad smells, hygiene problems and eutrophication, will be caused.

2.2.2. Industrial wastewater

Industrial wastewaters contain organic matter, nitrogen and phosphorus, according to the type of industry. Water quality of organic wastewater from industry is shown in

Table 2. It is obvious that BOD, T-N and T-P are high in wastewater from food and drug manufacturing factories.

Type of industry	BOD	T-N	T-P
Meat products manufacturing	300 - 600	50 - 80	10 - 15
Processed marine products manufacturing	200 - 2000	100 - 200	30 - 80
Seasoning manufacturing	300 - 200 0	100 - 150	15 - 60
Beer manufacturing	500 - 2000	30 - 50	5 - 15
Drug manufacturing	40 - 2500	80 - 100	10 - 20
Pulp manufacturing (kraft pulp)	300 - 700	150	2
Oil refining	20 - 200	20 - 30	5
Restaurant	30 - 3400	3 -42	1 - 12
Hospital	80 - 280	4 - 7	0.5 - 1

Table 2. Water quality of organic wastewater from industry

With the meat products manufacturing industry and processed marine products manufacturing industry, wastewater comes from treatment of raw material, boiling and cooling water. In brewing, wastewater comes from washing of malt, washing of machines, instruments and bottles, and cooling water. In the drug manufacturing industry, wastewater comes from the reaction, synthetic and distillation facilities.

Because these wastewaters contain high BOD, these wastewaters are normally treated by biological methods. If wastewater contains hazardous compounds, or if the pH is far from neutral, removal of hazardous substance and adjustment of pH are necessary before biological treatment.

2.2.3. Sewage effluent

In urban rivers in Japan, the amount of water from sewage works ranges from 40 to 80%. In Japan, the most popular wastewater treatment method is the standard activated sludge process. This achieves effective removal of organic matter, but the removal rate of nitrogen and phosphorus is not so high. Averaged water quality of treated wastewater in Japan is reported to be BOD 20 mg/l, T-N 25 mg/l and T-P 3 mg/l. In densely populated area, the amounts of nitrogen and phosphorus in treated water flowing into the sea, lakes and reservoirs are very high, leading to eutrophication. Nitrogen and phosphorus can be removed by advanced treatment, such as the Anaerobic/ Anoxic/ Oxic (A$_2$O) process. (Anaerobic means a condition without O$_2$, NO$_2^-$ and NO$_2^-$. Anoxic means condition with NO$_2^-$ and NO$_2^-$, without O$_2$.). In Japan the sewage effluent from only 8% of the population is subject to advanced treatment (having effluent quality less than BOD 10 mg/l, T-N 10mg/l and T-P 1 mg/l), but this is the best method of reducing eutrophication.

2.2.4. Livestock industry

Pollution from livestock farming contains high concentrations of organic matter, nitrogen and phosphorus. Of the total pollution load to Lake Kasumigaura in Japan, livestock farming contributes 12% of the COD, 13% of the nitrogen, and 5% of the phosphorus. Livestock feces from comprises 20% of total industrial waste in Japan.

The livestock industry can be classified into dairy cattle, beef cattle, sheep, pigs and poultry. Wastewater comes from feces, urine, leftovers of feed and washing water. The pollution load from livestock farms depends on the degree of separation of feces. If feces are separated properly, the pollution load from the livestock industry is reduced. Pollution load from pigs and cattle is shown in Table 3. The BOD load from pigs is three times higher than that of humans, and cattle are 20 times higher than humans.

		BOD	Nitrogen	Phosphorus
Pigs	Feces	114	19	13.3
	Urine	18	18	1.4
	Total	132	37	14.7
Cattle	Feces	720	129	51
	Urine	80	160	3
	Total	800	289	54

Source: Kai-Qin Xu *et al.* (1997)

Table 3. Pollution loads from raising livestock.

The characteristics of wastewater from the livestock industry are as follows:

- The amount of pollutants is high.
- Most of pollution load is in feces.
- Pollutant concentration is high.
- Biological treatment is available.
- Nitrogen concentration is high.

Table 3 indicates that 88% of BOD and 90% of phosphorus originates from feces. For reducing pollution load by wastewater treatment, it is important to remove feces before treatment. This should be treated into compost and recycled as fertilizer on agricultural fields. The treatment method differs according to the management of feces and urine. Mixed feces and urine are treated under anaerobic conditions and made into liquid fertilizer. Urine is treated with a biological treatment process such as activated sludge. Raw feces and compost made from feces are utilized in fields as fertilizer.

2.2.5. Fish farming

Organic matter, nitrogen and phosphorus are main the pollutants produced by fish farms. There are two types of fish farm—net cage culture, and freshwater lake and pond farms. If the effluent from a fish-farming pond is not treated and is allowed to enter lakes reservoirs and the sea, it will result in organic pollution and eutrophication. In net cage culture, waste food and feces sink to the seabed and are decomposed by microorganisms. Consequently dissolved oxygen in the bottom layer decreases and nitrogen and phosphorus are released to the water, where they are available to algae.

In lake Kasumigaura, where carp have been cultured in net cage culture, the nitrogen and phosphorus load per ton of production has been calculated to be 31 to 37 kg and 10 to 17 kg respectively. (Recently, however, culture of carp has been banned in Lake Kasumigaura, because of an epidemic of herpes.) In order to reduce the nitrogen and phosphorus load to the sea, lakes and reservoirs, research was conducted into the quality of the fish food. Fish farms contribute 6% and 19% respectively of the total nitrogen and phosphorus loads in Lake Kasumigaura, so fish farms have a particular responsibility for the phosphorus load. It was reported that the COD of the inner bay is correlated with the scale of salmon farming.

2.2.6. Landfill leachate

If landfill sites are not managed properly, many kinds of substances can pollute the surrounding environment. Landfill sites normally receive incinerated sludge, animal feces, garbage, industrial waste and other debris. Rainwater soaks into the waste and can transport pollutants as landfill leachate, the composition of which differs with the kind of waste. It contains organic matter, ammonium, heavy metals and other harmful substances. The concentration of organic matter mainly depends on the organic matter content of the waste, the amount of rain and biodegradability. The quality of measured landfill leachate was reported as COD_{Mn} 220 to 458mg/l, BOD 17 to 51mg/l, NH_4-N 340 to 660 mg/l and T-N 403 to 660mg/l, based on three years of research. Many kinds of substance are detected in landfill leachate, including harmful substances such as phthalic acid ester and dioxane.

3. Pollution of heavy metals and inorganic compounds

3.1. Adverse effects of heavy metals and inorganic compounds

Cadmium, lead, chromium, arsenic and mercury are typical heavy metals which have threatened human life. In Japan, pollution-related disease epidemics have occurred as a consequence of exposure to heavy metals. As an example of acute poisoning, cadmium causes acute gastroenteritis with vomiting, diarrhea and stomach ache. As for subacute poisoning, people may suffer from headaches, muscular rheumatism, liver trouble and kidney trouble. In the case of chronic poisoning, dry throat and nasal inflammation are often the first symptoms. Next may be yellow discoloration of the teeth, followed by sharp pains in the pelvis, the spine, and the distal bones of the limbs, and anemia. Finally, osteomalacia—typically with cracks in the shoulder blade and limb bones—arise and these bones are easily broken by a little shock. In Japan, there was an instance of water contaminated by cadmium from mining flowing into a river and paddy fields, contaminating rice and drinking water with cadmium. People who ate the rice and drank the water suffered chronic poisoning and osteomalacia.

Acute poisoning by inorganic mercury compounds causes oral pharyngitis, stomach ache, vomiting, turning pale, hypotension and increase of pulse rate. As for subacute poisoning, weariness of the whole body, loss of appetite, headache, mental disorder and central nerve disorder are caused. Chronic poisoning initially causes nervous breakdown such as depression, headache, dizziness, nervousness and memory impairment, followed by digestion disorders, insomnia, trembling of limbs and decrease of body weight.

The symptoms of acute poisoning by organic mercury compounds are similar to those of inorganic mercury compounds. They include exhaustion, loss of memory and attentiveness, paralysis of limbs, speech disorders, ataxia, and sight and hearing impairment.

In Minamata, Japan, an organic mercury compound was discharged into a coastal area from a chemical factory producing acetaldehyde. Organic mercury compounds accumulate in fishery products by bioconcentration, and people who ate contaminated fishery products suffered from chronic mercury poisoning. They mainly suffered disorders of the central nervous system; there were adverse effects on unborn children, and many people died.

Arsenious acid has high toxicity and is readily assimilated into the body. It causes acute poisoning with symptoms such as stomach cramps, vomiting, cholera-like diarrhea, dehydration, bad breath and reduced urination. In serious cases, patient shows symptoms of gastroenteritis, depressed temperature and blood pressure due to paralysis of arterioles and capillaries, followed by convulsions, coma and death. Symptoms of chronic poisoning include a weak constitution, loss of appetite, neuritis of lower limbs, keratosis, and hypertrophy and cirrhosis of the liver. Arsenic pollution occurred in Japan from the 1920s to the 1960s. Arsenic compounds were used as insecticides, weed killers and chemical weapons. Arsenopyrite was roasted to extract arsenious acid and this caused cinders and smoke containing arsenic compounds to be emitted, causing contamination of soil and crops around the mine. After some time, it became clear that the health of humans and livestock were clearly being affected by arsenic poisoning. In some parts of Japan, arsenic compounds can be detected in groundwater. This can occur naturally from geological sources, but much of this groundwater pollution in Japan originated from chemical weapons buried by the military.

Lead is used as an acid-resisting material because of its low reactivity. In the past, lead was used in women's face powder and lead poisoning was observed among women using it. It caused acute and subacute poisoning; the symptoms include a pale face, increase of pulse rate, vomiting, diarrhea and bloody stool. Exceptionally, there may be paralysis of the extensor in fingers and arms, and colic around the navel. As for chronic poisoning, the initial symptoms are weariness, loss of appetite, decrease of body weight, anemia, and discoloration of gums. If the poisoning get worse, there may be diarrhea due to disorder of digestive organs, constipation, stomach ache, arthralgia, paralysis of the extensors of the arms and trembling of fingers due to disorder of peripheral nerves. In serious cases, headache, dizziness, ringing in the ears, delusion, unconsciousness and sometimes blindness are caused due to disorder of the central nerve system. Lead can move to the unborn baby through the placenta, causing many pregnant women to miscarry or have a stillbirth.

Chromium is widely used in alloys and plating because of its hardness and heat-resistance, but chromium and its compounds also affect human health. The toxicity of chromium depends on its valence, hexavalent chromium compounds being the most toxic. If hexavalent chromium is in contact with skin, it reacts with proteins to form chromium compounds which eat into the skin. If a person assimilates hexavalent chromium orally, vomiting and diarrhea are caused and most of the chromium is excreted. The remainder is assimilated by digestive organs, leading to inflammation of kidneys and uremia. In cases of chronic poisoning, hexavalent chromium accumulates in the spleen, stomach, kidney and liver, acting as a cell poison. Furthermore, it has been proved to be a carcinogen.

There are many kinds of cyanide compounds, one of the most toxic being potassium cyanide. If cyanide compounds are assimilated orally, hydrogen cyanide is generated in the stomach. Hydrogen cyanide obstructs cell respiration and the brain is easily affected. The symptoms include headache, dizziness, ringing in the ears, consciousness disorder, increase of pulse rate and convulsions. In serious cases, enlargement of the pupil of the eye and decrease of body weight are observed, and the patient may die 15 to 45 minutes after assimilation of cyanide compounds. As for subacute poisoning, it is said that chronic exhaustion, headache, dizziness, loss of appetite, mental derangement and general metabolic disorder are observed.

Metals such as manganese, aluminum, copper and iron very rarely cause human poisoning, but they often cause discoloration of water. This causes problems when it is used for domestic purposes such as washing. These metals elute from the bottom of water bodies when dissolved oxygen in the bottom layer is depleted. If water in the public supply is colored by metals, they must be removed in the water treatment works.

Nitrate and nitrite cause methemoglobinemia, or 'blue baby disease', and nitrite causes production of carcinogenic N-nitroso compounds in the human stomach. In saliva and the stomach, there are nitrate deoxidizing bacteria which produce nitrite. If an adult drinks nitrate-contaminated water, nitrate deoxidization is retarded because the pH of the stomach is approximately 2. In the case of babies, however, the pH is over 4, and nitrate deoxidization does occur, producing nitrite compounds. These react with hemoglobin and methemoglobin is produced in the blood. This does not have the oxygen-carrying capacity of hemoglobin. If the methemoglobin to hemoglobin ratio exceeds 10%, cyanosis is caused and sometimes the patient dies.

In Japanese drinking water quality standards, the total concentration of nitrate and nitrite is regulated to less than 10 mg/l as nitrogen. In Europe and USA, pollution of drinking water by nitrate and nitrite is a serious problem. European countries depend on groundwater for potable supplies and management and regulation of nitrate and nitrite in agricultural fields have become important countermeasures.

3.2. Pollution source of heavy metals and inorganic compounds

The effluent from industrial facilities is likely to contain pollutants that pose a risk to human health if the company does not manage its waste water properly. The pollutants vary according to the type of industry. Effluent from the food industry contains mainly organic matter, nitrogen and phosphorus, which is treated by biological wastewater treatment. Heavy metals are emitted from mining, the non-ferrous metal refining industry, plating factories, the chemical industry, and leather and electronic manufacturers. From mining, copper, arsenic and cadmium had been reported to cause pollution and human poisoning. The effluent from plating factories potentially contains zinc, copper, nickel and chromium. Effluent from electronics manufacturers potentially contains lead from solder, but it is usually controlled. Chromium is used for tanning in leather manufacturing. Cyanide compounds are contained in wastewater emitted from plating factories, smelters and coke producers. Wastewater containing heavy metals is generally treated by a chemical process such as coagulation-precipitation and ion exchange.

Nitrate and nitrite pollution is mainly caused by fertilization of agricultural fields. Ammonium sulfate and urea are widely used as fertilizers. Unutilized ammonium, nitrite and nitrate produced by nitrification move into groundwater, streams, rivers and lakes and cause nitrogen pollution of drinking water and eutrophication.

4. Pollution of harmful organic compounds

4. 1. Adverse effect and pollution source of organochlorine compounds

Typical organochlorine compounds are trichloroethylene and tetrachloroethylene. Greases and oils are readily dissolved by these compounds, so they are used for cleaning of metal parts and electronics, as well as for dry cleaning and in paint. If such chemicals are not managed properly, pollution of soil and groundwater can occur. General characteristics of organochlorine compounds are that they are refractory and have low viscosity, low surface tension, and high volatility and persistency. If organochlorine compounds are allowed to penetrate into soil, their concentration can increase to around the saturated solubility if the local flow rate of groundwater is slow and the contact time between the compounds and groundwater is long. Organochlorine compounds cause contamination of groundwater and are regulated by effluent standards and environmental quality standards. Assimilation of trichloroethylene causes depression of the central nervous system, vomiting, and stomach ache. Assimilation of tetrachloroethylene causes liver dysfunction, unconsciousness, vomiting and stomach ache. Tetrachloroehylene is suspected of being a carcinogen (IARC group 2B, possibly carcinogenic to humans).

Polychlorinated biphenyls (PCBs) have been used in insulators for electrical facilities, lubricating oil and plasticizers. PCB was proved to cause cancer, skin disorders, and ailments of the liver and digestive organs. Production of PCB was banned in Japan, in 1970. However, residual PCB was detected in the environment because the amount of PCB used was very high, and microbial degradation of PCB in the environment is very slow.

4.2. Adverse effects and pollution sources of agricultural chemicals

Agricultural chemicals are mainly classified into insecticides, disinfectants, weed killers and rat poisons. Many of them are organic substances and are classified by chemical formulation into organophosphorus compounds, organochlorine compounds, organosulfur compounds, triazine compounds, diphenylether compounds, and organotin compounds. Agricultural chemicals in the past such as DDT (dichloro diphenyl trichloroethane) had high acute toxicity and long duration of toxicity in the environment. Poisoning symptom of humans exposed to organophosphorus compounds are optical disorders, vomiting, diarrhea, dyspnea and paralysis of the central nervous system.

Human poisoning by organophosphorus compounds rapidly leads to death. Poisoning symptom of humans exposed to organochloride compounds are headache, nausea, dizziness, trembling, ataxy, and convulsion.

Organochloride pesticides such as DDT are readily assimilated into the body of living organisms and are only very slowly excreted. People can be adversely affected by these agricultural chemicals through biological accumulation. Poisoning symptoms of animals exposed to triazine compounds are increase of mortality, loss of appetite and decrease of body weight. Puffiness of the lungs and kidneys, and bleeding of the surface of the heart may be observed.

Highly toxic or persistent agricultural chemicals such as DDT are now neither used nor registered for sale in Japan owing to both the strengthening of regulations regarding their use under the Agricultural Chemicals Regulation Law, revised in 1971, and the development of less toxic substitutes by advanced science and technology. Therefore, the environmental pollution currently caused by persistent agricultural chemicals has decreased. However, soil and groundwater are still polluted by agricultural chemicals. Pollution of soil and groundwater occurs as a result of inappropriate application and management of agricultural chemicals.

4.3. Adverse effects and sources of dioxins

	%
General waste incinerator	46
Industrial waste incinerator	33
Small incinerator	11 - 12
Manufacturing process	9
Effluent treatment facility	0.02
Others (cigarettes, exhaust gas from automobiles)	0.1 - 0.7

Source: Ministry of the Environment (2001)

Table 4. Sources of dioxin in Japan, and the percentage of the whole contributed by each.

Of all the organic compounds, it is said that dioxin is the most toxic. About 95% of the dioxins in the atmosphere is supposed to be emitted from incinerators. Dioxins include polychlorinated dibenzo-para-dioxin (PCDD), polychlorinated dibenzo furan (PCDF) and coplanar-polychlorinated biphenyl. According to the substitution number and locus of chlorine atoms, there are 75 PCDD's and 135 PCDF's. The toxicities of homologues differ and 2,3,7,8-TCDD is the most toxic compound. Dioxins are discharged from general waste incinerators, industrial waste incinerators, small incinerators, manufacturing processes, effluent treatment facilities and other sources (e.g. cigarettes and exhaust gases from automobiles) (see Table 4). Dioxin is discharged from municipal or industrial incinerators during the combustion process. Atmospheric dioxins come down to the soil and water environment with rain and dust. Pollution of soil by dioxin can cause accumulation of dioxin in farm products. Pollution of water by dioxin can cause health problems through water ingestion and eating fishery products. Dioxin poisoning reportedly causes teratogenetic foetuses, cancer, atrophy of the thymus, and reproductive disorders.

In Japan, environmental standards for dioxin in soil and water have been established, at 1000 pg-TEQ/g for soil and 1 pg-TEQ/l for water. In 2002, 3300 soil samples were analyzed and no sample exceeded the standard for dioxin, 1976 water samples were analyzed and 56 samples exceeded the standard. Averages dioxin concentrations were 3.8 pg-TEQ/g for soil (range: 0 to 250) and 0.25 pg-TEQ/l (range: 0.01 to 2.7) for water.

4.4. Adverse effects and sources of endocrine disruptors

There are harmful substances in the environment that can affect the endocrine system of animals after absorption to the body. These are called endocrine disruptors. Chemical compounds such as DDT, PCB, nonylphenol and tributhyltin compounds have been proved to be endocrine disruptors. Endocrine disruptors have caused masculinization and feminization of molluscs, fish, crustaceans, birds and mammals. This has led to reduced hatching and populations of some species around the world. It was reported male genitalia formed in female periwinkles, as a result of masculinization by an organotin compound. It was also reported that male rainbow trout can be feminized by nonylphenol. Organochlorine agricultural chemicals such as DDT are believed to cause dwarfism of the penis of alligators. PCB is considered to cause loss of immunity population decline in seals. There is also evidence that estrogen in sewage effluent can cause feminization of fish in urban rivers. Masculinization and feminization cause reproductive disorders and population decline in aquatic organisms.

Endocrine disruptors have been discharged principally by industry. In the aquatic environment in Japan, the most frequently detected endocrine disruptors are PCB, nonylphenol, 4-tert-octylphenol, bis-phenol A and estrogen. Organotin compounds such as tributyltin and triphenyltin compounds are used as antifoulant paints on the hull of ships. In urban rivers, estrogen from excreta has been frequently detected. In drinking water resource, di-2-ethylhexyl phthalate (DEHP), di-n-buthyl phthalate (DBP), bis-phenol A (BPA) and phenol are frequently detected. BPA is used as a component of plastics. DBP and DEHP are used as plasticizers. Nonylphenol is used as a surfactant and synthesis material for phenolic resins and esters. PCB had been used for insulation.

Results with removal of endocrine disruptors, nonylphenol, BPA and 17β-estradiol as estrogen in wastewater treatment process have been reported. The removal rate of nonylphenol was high irrespective of the hydraulic retention time (HRT), but those of BPA and 17β-estradiol were low at low HRT. The removal rate of BPA and 17β-estradiol increased with increase of HRT. Thus the biodegradabilities of endocrine disruptors are different from each other. To treat wastewater containing endocrine disruptors, the HRT of the aeration tank should be enough for treatment of endocrine disruptors.

5. Countermeasures against water pollution

Water bodies are utilized for various purposes, such as drinking water, agriculture, industry, fisheries and recreation. To protect human health and satisfy water quality for each purpose, standards of environmental quality and effluent standards are necessary.

In many countries now, environmental quality standards and effluent standards have been established for control of water pollution of lakes, reservoirs, rivers and coastal water.

Effluent standards regulate wastewater discharged from factories and business establishments. In Japan, national effluent standards have two categories—items related to the protection of human health and those related to the conservation of natural environment (see Table 5). For items related to the protection of human health, permissible limits of heavy metals, organochlorine compounds, and agricultural chemicals are established. For conservation of the natural environment, permissible limits of water parameters such as BOD, COD, nitrogen and phosphorus are established.

Toxic substances	Permissible limits
Cadmium and its compounds	0.1 mg/l
Cyanide	1 mg/l
Organophosphorus compounds (only parathion, methyl parathion, methyl demeton)	1 mg/l
Lead and its compounds	0.1 mg/l
Hexavalent chrome compounds	0.5 mg/l
Arsenic and its compounds	0.1 mg/l
Total mercury	0.005 mg/l
Alkylmercuric compounds	Not detectable
PCB	0.003 mg/l
Trichloroethylene	0.3 mg/l
Tetrachloroethylene	0.1 mg/l
Dichloromethane	0.2 mg/l
Carbon tetrachloride	0.02 mg/l
1,2-Dichloroethane	0.04 mg/l
1,1-Dichloroethylene	0.2 mg/l
cis-1,2-Dichloroethylene	0.4 mg/l
1,1,1-Trichloroethane	3 mg/l
1,1,2-Trichloroethane	0.06 mg/l
1,3-Dichloropropene	0.02 mg/l
Thiram	0.06 mg/l
Simazine	0.03 mg/l
Thiobencarb	0.2 mg/l
Benzene	0.1 mg/l
Selenium and its compounds	0.1 mg/l
Boron and its compounds	Non-marine 10 mg/l, marine 230 mg/l

Flouride and its compounds	Non-marine 8 mg/l, marine 15 mg/l
ammonia, ammonium compounds, nitrate compounds and nitrite compounds	100 mg/l*

*: A sum total of ammonia-N times 0.4, nitrate-N and nitrite-N

A. Items related to the protection of human health

Living environment items	Permissible limits
pH	Non-marine5.8-8.6, marine 5-9
Biochemical Oxygen Demand (river)	160 mg/l(Daily average120 mg/l)
Chemical Oxygen Demand (lake, reservoir and coastal area)Oxidizer: potassium permanganate	160 mg/l(Daily average120 mg/l)
Suspended Solid	200 mg/l(Daily average150 mg/l)
n-hexane extracts(mineral oil)	5 mg/l
n-hexane extracts(animal oil and vegitable oil)	30 mg/l
Phenolic compounds	5 mg/l
Copper	3 mg/l
Zinc	5 mg/l
Dissolved iron	10 mg/l
Dissolved manganese	10 mg/l
Chromium	2 mg/l
Coliform bacteria	Daily average 3000 colonies/cm^3
Nitrogen	120 mg/l(Daily average 60 mg/l)
Phosphorus	16 mg/l(Daily average 8 mg/l)

B. Items related to the conservation of living environment

Table 5. Effluent standards set by the Water Pollution Control Law in Japan

Environmental quality standards for water pollutants have been established to protect human health and conserve the natural environment in Japan. Standards for water pollutants also have two categories—environmental quality standards for human health and those related to the conservation of the natural environment. In environmental quality standards for human health, standard values have been established for heavy metals, organochlorine compounds, and agricultural chemicals. For environmental quality standards related to the natural environment, rivers, lakes, reservoirs and coastal waters are classified according to their use, such as drinking water, agriculture, industry and fisheries. Standard values of BOD, COD, pH, SS, DO, total coliforms, nitrogen and phosphorus are established for each class.

Prefectural governors must check the quality of water bodies periodically. At the same time, factories and business establishments must analyze the quality of their wastewater and evaluate compliance with the national effluent standards (Table 5). Where it is judged that the national effluent standards is insufficient to attain the environmental quality standard in a certain water body, the prefectural governor is authorized to introduce a more stringent prefectural standard through a prefectural ordinance.

Since pollution by dioxin from waste incinerators became a serious problem in Japan, the Special Control Law for Dioxins was enacted in 2000, aimed at prevention of environmental pollution by dioxin, elimination of dioxin and protection of human health. This law defines dioxins as polychlorinated dibenzo-para-dioxin (PCDD), polychlorinated dibenzofuran (PCDF) and coplanar polychlorinated biphenyl (coplanar PCB). This law sets the tolerable daily intake at 4 pg-TEQ or less per 1 kg of body weight.. The environmental standards for dioxin in water and soil are based on this value. By law, the quality of dioxins in effluent from facilities was established as 10 pg-TEQ/l, without reference to the kind of facility.

In order to prevent pollution by agricultural chemicals in Japan, the law requires that any agricultural chemical proposed for use must be registered for sale, a process which includes evaluation of the persistence, toxicity, and other characteristics. All agricultural chemicals registered for sale must first satisfy criteria in each of the following categories: (1) persistence in agricultural products; (2) persistence in soil; (3) toxicity to aquatic animals and plants; (4) water pollution. As of March 2003, the standards for persistence in agricultural products have been established for 383 agricultural chemicals and the standards for water pollution have been established for 135 chemicals. A common standard for persistency in soil and toxicity against aquatic animals and plants has been established for all agricultural chemicals.

6. Other forms of pollution

6.1. Oceanic pollution

Pollution of the ocean includes mainly oil pollution, drifting wastes and red tides. In 2003, 571 cases of ocean pollution were reported in Japan. Oil pollution comprised approximately 70% of the cases. Most oil pollutions is caused by carelessness and accidents in ships. Most of the drifting wastes were petrochemicals such as styrene foam and vinyl. Marine animals accidentally swallow drifting waste, significantly jeopardizing their health.

It is difficult to evaluate the pollution status of the World Ocean because research is mainly limited to developed countries. In enclosed water bodies such as the North Sea, Baltic Sea and Mediterranean Sea, the extent of red tides and of heavy metal pollution are increasing.

Oil spills occur in connection with offshore oil production and loading of tankers. Oil spills due to shipping accident often have serious environmental effects because the oil can spread over a very wide area and the duration of the pollution is long because oil generally contains refractory organic compounds. In 1997, there were two oil spills as a result of shipping accidents in Japan. In these cases, oil drifting ashore was collected quickly, but serious environmental damage still occurred.

6.2. Emission of SOx and NOx causing acid rain

Acid rain is a phenomenon in which nitrate and sulfate is present in precipitation, and generally is defined as rain with pH less than 5.6. Nitrate and sulfate are formed by chemical reactions of NOx and SOx generating from combustion of fossil fuel.

Soot and smoke from factories contains NOx and SOx. These gasses cause acid rain and decrease the pH of rivers, lakes and reservoirs. Decrease of pH effects aquatic life such as plants and fish. Emission of NOx from factories increases the nitrogen concentration of rain, which increases nitrogen load to rivers, lakes and reservoirs and promotes eutrophication.

Acid rain disturbs plant metabolism, decreases the pH of soil and promotes elution of aluminum and zinc. Acid rain also decreases the pH of lakes and reservoirs, and affect aquatic life. In northern Europe, where acid rain is a serious problem, lime is introduced to soil and lakes to prevent decrease of pH. Because NOx and SOx move between countries, cooperative monitoring of acid rain and international countermeasure are necessary.

NOx and SOx are emitted from various industries, transportation and automobiles that consume fossil fuel. To prevent acid rain, it is important to control emission of SOx and NOx. For reduction of SOx emission, it is necessary to remove sulfur from fuel and to extract the SOx from smoke emitted from factories and power stations. Measures for reduction of NOx emission include improvements in combustion such as lowering of combustion temperature, and removal of NOx in smoke from factories. Increased utilization of fuel in which nitrogen content is low, such as natural gas, is also important. For reduction of NOx from automobiles, standards of NOx emission have been established in Europe, Japan and USA.

6.3. Thermal pollution

Increase of water temperature in public water bodies is regarded as a form of anthropogenic water pollution. Increase of temperature is caused by discharge from thermal power plants, atomic power plants, iron manufacturing and the chemical industry, but most thermal pollution is attributable to power plants. Sea water is used in condensers for cooling steam from steam turbines in coastal power plants. The temperature of the seawater increases 7 to 10 degrees in the condenser before it is discharged to the sea. A power plant generating 350 000 kW of electricity discharges 0.86 to 1.7 million m^3/day of warm water.

Thermal discharge generally spreads over the surface layer from the discharge point and the water temperature increases by 1 °C within a radius of approximately 700 to 1000 m.

Thermal discharge is considered to affect fisheries. All forms of marine life have an optimum temperature, and increase of temperature by thermal discharge may cause extermination of native marine organisms and colonization by other species. Feeding grounds for migratory fish can be affected and migration routes may be changed. It is known that growth of wakame seaweed is inhibited over 25 °C, indicating that this fishery may be damaged by thermal discharge.

Glossary

A$_2$O process:	A modification of the activated sludge process for enhanced nitrogen and phosphorus removal. It involves an anaerobic tank (without dissolved or combinative oxygen), an anoxic tank (without dissolved but with combinative oxygen, e.g. NO$_3^-$, NO$_2$), an aeration tank and a sedimentation tank. In the anaerobic tank, phosphorus accumulating bacteria proliferate. In the anoxic tank, denitrification by facultative anaerobic bacteria proceeds. In the aeration tank, nitrification by nitrifying bacteria and phosphate absorption by phosphorus accumulating bacteria proceed. Some of the activated sludge forming in the aeration tank is returned to the anoxic tank for denitrification.
Activated sludge process:	A common wastewater treatment process consisting of an aeration tank and a sedimentation tank. In the aeration tank, microorganisms such as bacteria and fungi assimilate organic matter and proliferate in aerobic conditions, and protozoa and metazoa predate bacteria and fungi and proliferate, forming 'activated sludge'. The activated sludge then flows into the sedimentation tank, followed by settling of microorganisms to produce clear water. The clear water is discharged to the aquatic environment after disinfection. Settled microorganisms are returned to the aeration tank or are incinerated after digestion treatment and dehydration.
Algal bloom:	A phenomenon in which algae accumulate on the surface of lakes, reservoirs and the sea. The surface water is colored green, red or brown (recently, toxic substances produced by cyanobacteria have become a world wide environmental issue).
BOD (Biochemical Oxygen Demand):	An index of pollution by organic matter. Consumption of oxygen by respiration of aerobic microorganisms assimilating organic matter in water under constant conditions.
BPA:	Bis-phenol A
Coagulation-sedimentation process:	A water purification system introduced in waterworks for removal of suspended solids. In a mixing pond, aluminum sulfate is mixed with raw water as a coagulant. Suspended solids flocculate in the flocculation pond and settle in the settling pond.
COD$_{Mn}$:	Chemical Oxygen Demand using potassium permanganate as oxidizer. An indicator of pollution by organic matter.
DBP:	Di-N-Buthyl Phthalate.
DDT:	Dichloro Diphenyl Trichloroethane.
DEHP:	Di-2-Ethylhexyl Phthalate.
Eutrophication:	Inflow of nitrogen and phosphorus from human activities increases

	concentrations of nitrogen and phosphorus in lakes, reservoirs, and inland sea. Primary production increases and algal blooms and red tides occur.
Fertilizer:	Artificially spread nutrient for crops, e.g. ammonium sulfate.
Filter:	A basin filled with sand for removal of suspended solids after the coagulation-sedimentation process.
HRT (Hydraulic Retention Time):	An index of water exchange. For an aeration tank in the activated sludge process, HRT (h) is calculated as V/Q. Q: amount of wastewater flowing into aeration tank per hour (m^3/h), V: volume of aeration tank (m^3).
IARC:	International Agency for Research on Cancer
Landfill leachate:	Water percolated from landfill, containing pollutants originated from wastes.
NH_4-N:	Ammonium nitrogen.
NOx:	Generic of nitrogen oxide. The main nitrogen oxides causing air pollution are nitric oxide (NO) and nitrogen dioxide(NO_2).
PCB:	Polychlorinated Biphenyl.
PCDD:	Polychlorinated Dibenzo-Para-Dioxin.
PCDF:	Polychlorinated Dibenzo Furan.
Pre-chlorination:	Pre-chlorination is introduced in waterworks for treatment of heavily polluted water. Polluted water is dosed with sodium hypochlorite or liquid chlorine for removal of organic matter, ammonia, iron, manganese and bacteria.
Sewage works:	A facility removing organic matter, nitrogen and phosphorus from wastewater and discharging treated water to rivers, lakes and the sea.
SOx:	Generic of sulfur oxide. The main sulfur oxide causing air pollution are sulfur dioxide (SO_2) and sulfur trioxide (SO_3).
SS (Suspended Solids):	Dry weight of suspended solids per volume of water. In the case of natural water, aquatic organisms such as bacteria, algae and zooplankton, and soil particles are analyzed as SS.
TCDD:	Tetrachlorinated Dibenzo-Para-Dioxin.
TEQ:	Toxic Equivalent.
T-N:	Total Nitrogen.
T-P:	Total Phosphorus.
Waterworks:	An entire facility distributing tap water to people, including a water purification plant.
WHO:	World Health Organisation.

Bibliography

Foundation of River and Watershed Environment Management, Effect of nutrient concentration on the water quality of river, in Japanese.

Japan environmental management for industry, 'Kougaiboushi no gijutsu to houki' (1995) in Japanese.

Kai-Qin Xu, Hye-Ok Chun, Ryuichi Sudo, Journal of Water and Waste, 39, 1097-1105 (1997), in Japanese.

Ministry of the Environment, Quality of the Environment in Japan(2004) in Japanese.

Ministry of the Environment, 'Suisitsuodaku' the first volume(1973) in Japanese.

Ministry of the Environment, Water Environment Management in Japan (2001).

Munehiro Nomura, Nobuo Chiba, Kai-Qin XU and Ryuichi Sudo, The effect of pollutant loading from the fishery cultivation on water quality in inner bay, Journal of Japan Society on Water Environment, 21, 719-726 (1998) in Japanese.

Myung-Sook Jung, Kai-Qin Xu, Yuhei Inamori, Masaaki Hosomi and Ryuichi Sudo, Evaluation of treatment characteristics of landfill leachate and effect on the aquatic ecosystem using microtox tests, Journal of Japan Society on Water Environment, 19, 922-929 (1996) in Japanese.

Sakae Kuribayashi *et al*, Nationwide survey of endocrine disrupting chemicals in wastewater treatment plants in Japan, Journal of Water and Waste, 44, 39-45 (2002) in Japanese.

Toshio Takeuchi, Reduction of nitrogen and phosphorus from freshwater fish farm, Journal of Water and Waste, 44, 611-619 (2002) in Japanese.

Biographical Sketches

Yuhei Inamori is an executive researcher at the National Institute for Environmental Studies (NIES), where he has been in his present post since 1990. He received B.S. and M.S. degrees from Kagoshima University, Kagoshima Japan, in 1971 and 1973 respectively. He received a PhD degree from Tohoku University, Miyagi, Japan in 1979.

From 1973 to 1979, he was a researcher for Meidensha Corp. From 1980 to 1984, he was a researcher for NIES, and from 1985 to 1990, a senior researcher for NIES. His fields of specification are microbiology, biotechnology and ecological engineering. His current research interests include restoration of aquatic environments using bioengineering and eco-engineering applicable to developing countries. He received an Award of Excellent Paper, from Japan Sewage Works Association in 1987, an Award of Excellent Publication from the Journal "MIZU" in 1991, and an Award of Excellent Paper, from the Japanese Society of Water Treatment Biology in 1998.

Naoshi Fujimoto is Assistant Professor of Tokyo University of Agriculture, where he has been at his present post since 2000. He received B.E. and M.E. degrees from Tohoku University, Miyagi, Japan, in 1991 and 1993 respectively. He received a D.E. degree in civil engineering from Tohoku University in 1996. From 1996 to 1999, he was a research associate of Tokyo University of Agriculture. His specialist field is environmental engineering. His current research interests include control of cyanobacteria and its toxins in lakes and reservoirs.

POINT SOURCES OF POLLUTION

Yuhei Inamori

National Institute for Environmental Studies, Tsukuba, Japan, and

Naoshi Fujimoto
Faculty of Applied Bioscience, Tokyo University of Agriculture, Tokyo, Japan

Keywords: organic matter, nitrogen, phosphorus, heavy metals, domestic wastewater, food industry, chemical industry, activated sludge process, biofilm process, anaerobic treatment

Contents

1. Introduction
2. Kind of point sources
3. Countermeasures for point sources
4. Wastewater treatment processes
Glossary
Bibliography
Biographical Sketches

Summary

Point source loading refers to pollutants produced by identified pollution sources. Point source loadings include treated and untreated wastewater from factories, domestic wastewater, treated water from sewerage treatment plants, and wastewater from feedlots and fish farms. Organic matter, nitrogen and phosphorus are discharged into water bodies by inflow of domestic wastewater and industrial wastewater causing organic pollution and eutrophication.

The point source loading of organic matter, nitrogen, and phosphorus as percentages of the total loading in Lake Kasumigaura in Japan are 55%, 58%, and 77% respectively. Among point source loadings, the largest loading is domestic wastewater, which supplies 33% of COD_{Mn}, 34% of nitrogen and 45% of phosphorus.

In Japan, the amount of wastewater, BOD, nitrogen and phosphorus were calculated approximately to 200 L, 40g, 10g, and 1g, respectively, per person per day. Averaged water quality of domestic wastewater is BOD 200 mg/l, total nitrogen (T-N) 50 mg/l and total phosphorus (T-P) 5 mg/l.

The concentrations of organic matter, nitrogen, and phosphorus in wastewater from food factories are generally high and each vary widely according to the factories producing them. For reduction of pollution from food factories, it is important reduce output of pollutants from each stage of production, e.g. pretreatment, manufacturing, finalization, machine and instrument washing, and wastewater treatment. Separation of solid waste and wastewater, and water saving are also important.

Wastewater from leather manufacturing factories and electroplating factories can contain heavy metals and organochloride compounds. Wastewater from electroplating factories can be highly toxic because it can contain hexavalent chromium and trichloroethylene.

Treatment of wastewater at source, provision of sewerage, construction of small scale wastewater treatment plants in regions where it is difficult to construct sewage works, construction of other kinds of industrial wastewater purification facilities and septic tanks (on-site domestic wastewater treatment facilities called JOHKASOU), are all important for control of point source pollution.

Because wastewater from kitchens accounts for a high percentage of the pollutants in domestic wastewater, and contains large quantities of organic matter, nitrogen, phosphorus and suspended material, measures to deal with wastewater must be implemented in order to conserve the aquatic environment.

1. Introduction

Many kinds of pollutants affect the living environment, human health and ecosystems. Pollutants such as organic matter, nitrogen, phosphorus, heavy metals, organochlorine compounds, dioxins and agricultural chemicals, etc. are discharged to surface water bodies by human activity. Increase of organic matter in water bodies causes depletion of dissolved oxygen, proliferation of pathogens, alteration of environmental conditions for aquatic life, offensive odors and landscape deterioration. Increase of nitrogen and phosphorus causes eutrophication, which leads to many problems such as toxic algal blooms, blockages in water works, and depletion of dissolved oxygen in the bottom layer. Most of the organic matter, nitrogen and phosphorus are discharged into water bodies in domestic and industrial wastewater. Harmful compounds affecting human health are discharged by the chemical industry, landfills, agricultural fields and incinerators.

Pollution sources are categorized as point source and non-point source. The former refers to pollutants discharged from identified pollution sources. Non-point source refers to pollutants discharged from a wide area. Because the location of a point source can be readily identified, it is possible to control its pollution loading. Point source loadings include domestic and industrial wastewater, treated water from sewage and industrial wastewater treatment plants and Johkasous, feedlots and fish farms. Pollutants discharged from point sources include organic matter, nitrogen, phosphorus, heavy metals, and organochlorine compounds, etc. In this chapter we provide an outline of point source loading, typical kinds of point sources and their water quality, countermeasures for point source loading, and an introduction to wastewater treatment processes.

2. Kind of point sources

2.1. Percentage of point source loading to total pollution loading

The point source loading of organic matter (COD_{Mn}),, nitrogen, and phosphorus as percentages of the total loading in Lake Kasumigaura in Japan are 56%, 58%, and 77% respectively. Note that the point source loading of phosphorus is particularly high (see Figure 1). A breakdown of point source loading reveals that the largest loading is domestic wastewater, which supplies 33% of COD_{Mn}, 34% of nitrogen and 45% of the phosphorus. The second largest source of COD_{Mn} and nitrogen is livestock wastewater, which supplies 10% of these substances. Fisheries wastewater supplies about 20% of the phosphorus. The point source loading as a percentage of the total pollution loading in Lake Biwa is organic matter (COD_{Mn}) 53.1%, nitrogen 45.3%, and phosphorus 64.8%, revealing that, as in Lake Kasumigaura, the percentage of phosphorus loading from point sources is highest.

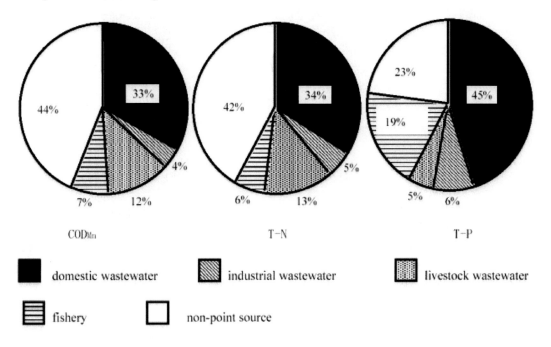

Figure 1. Percentage of pollution sources of Lake Kasumigaura

Percentages of pollution loadings to typical enclosed coastal seas in Japan from industrial and domestic wastewater are shown in Figure 2. Pollution loading of industrial wastewater to Tokyo Bay, Ise Bay, Seto Inland Sea and three other sea areas account for 21%, 34%, 41% and 35% respectively. Pollution loading of domestic wastewater account for more than 50% in three sea areas.

2.2. Domestic wastewater

The point source with the largest pollutant loading is domestic wastewater. The unit quantities of organic matter, nitrogen, and phosphorus in gray water, after excluding feces and urine from domestic wastewater are: BOD, 27 g/person/day; nitrogen, 2 g/person/day; and total phosphorus, 0.4 g/person/day (see Figure 1 in *Pollution Sources*). The unit quantities of organic matter, nitrogen, and phosphorus in feces and urine are BOD 13 g/person/day, nitrogen 8 g/person/day, and phosphorus 0.6 g/person/day. This reveals large nitrogen and phosphorus loadings in feces and urine. Such point source loadings appear where domestic wastewater is discharged without treatment in regions with no sewage treatment systems or Johkasous. In the home, pollutants are generated mainly from excretion, cooking, washing and bathing. Major pollutants are organic matter, nitrogen and phosphorus. They come from cooking oil, detergent, body soap, and food, in addition to excrement. In Japan, the amount of wastewater, BOD, nitrogen and phosphorus were calculated approximately to 200 L, 40g, 10g, and 1 g per person per day. Averaged water quality of domestic wastewater is BOD 200 mg/l, T-N 50 mg/l and T-P 5 mg/l.

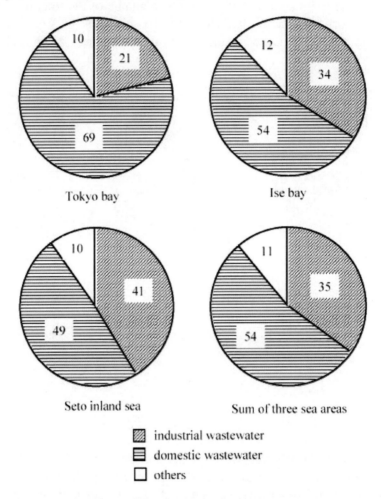

Figure 2. Percentage of COD loading from industrial and domestic wastewater in three typical enclosed coastal seas in Japan

Domestic wastewater is treated by biological treatment such as the activated sludge process and the biofilm process. Treatment facilities are classified into sewage works, rural community sewerage, and Johkasous installed in individual house. If domestic wastewater is treated effectively by sewage works, rural community sewerage and Johkasou, pollution loading from domestic water use can be greatly reduced. However, if domestic wastewater is not treated and flows directly to water bodies, the result is decrease of dissolved oxygen, bad smell, proliferation of pathogens and eutrophication. In the past, Johkasou treatment only of night soil was popularized in Japan, causing pollution of water bodies by the organic matter, nitrogen and phosphorus contained in gray water. Today, installation of Johkasous that only treat night soil is forbidden.

2.3. Industrial wastewater

2.3.1. Food industry

The concentrations of organic substances, nitrogen, and phosphorus in wastewater from food factories vary widely according to the industry producing them.

Wastewater from food manufacturing factories contains a high concentration of BOD, T-N and T-P in general (see Table 1). Wastewater from fish canning plants, and manufacturers of flavor enhancers, baking powder, yeast and starch, contains particularly high BOD, T-N and T-P. For control of organic pollution and eutrophication, removal of organic matter, nitrogen and phosphorus from food manufacturing factories is vital.

	BOD	T-N	T-P
Livestock products	600	50	15
Dairy products	250	35	5
Fish canning	2700	210	75
Flavor enhancers	1000	460	50
Sugar	450	25	5
Bakery	1300	30	15
Beer	1000	40	10
Baking powder, yeast	7000	450	30
Starch	2300	130	30

Table 1. Quality of wastewater from food factories

Wastewater from food manufacturing factories is generated from four sources: pretreatment, manufacturing, finalization and machine and instrument washing (Figure 3). Useless substance attaching to the raw material, unnecessary raw material, spoiled material and residual material in the machine and instruments, are contained in wastewater. For reduction of pollution from food factories, pollutants must be reduced in each process, and wastewater treatment is also important. In each process there should be separation of solid waste and wastewater, and water saving measures should be introduced.

A flow sheet for selection of wastewater treatment method is shown in Figure 4. This can be applied to all kind of industries. After removal of garbage and oil, wastewater is stored in a reservoir tank so that water quality and amount of wastewater can be kept constant. Investigation of the quality of wastewater can lead to selection of appropriate treatment methods, for inorganic, organic and hazardous wastewater. In general, wastewater from food factories is organic wastewater, which is treated by biological treatment. There are many biological treatment processes and managers of wastewater treatment plants must select the method according to the characteristics of the wastewater, the target for effluent quality, area available for the wastewater treatment facility and meteorological conditions.

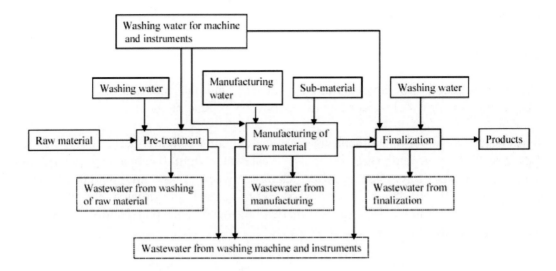

Figure 3. Sources of wastewater in food manufacturing factories

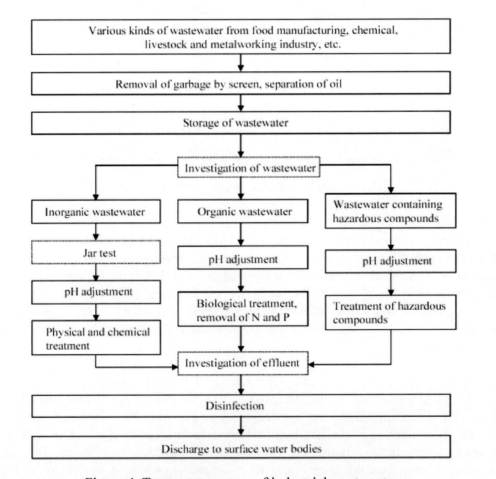

Figure 4. Treatment process of industrial wastewater

Wastewater treatment processes used by the food industry include activated sludge, biofilm processes such as trickling filter, rotating disc and submerged filter, and anaerobic biological treatment. The quality of wastewater and effluent from the treatment processes of various food factories are shown in Table 2. BOD exceeds 2000 mg/l for all types except for factories producing vegetable oils and spreads. The nitrogen concentration of wastewater is high at factories producing ham and bacon, lard for eating and flavor enhancers. The nitrogen concentration of wastewater from factories producing flavor enhancers exceeds 1000mg/l. For most factories, the phosphorus concentration of wastewater exceeds 50 mg/l and it is highest in the case of vegetable oils, at 150 mg/l. Most factories have introduced the activated sludge process. If wastewater contains oil, an oil separation tank is introduced before biological treatment. BOD concentration of effluents is reduced to less than 5 mg/l, except for vegetable oils. The nitrogen concentration of effluent is highest in the case of flavor enhancers. Phosphorus concentration of effluent from vegetable oil factories is high. Thus removal of organic matter is normally achieved, but removal of nitrogen and phosphorus is not achieved by some factories. Removal of nitrogen and phosphorus from the wastewater from food manufacturing factories is important in control of eutrophication.

Products	Treatment process	Quality of wastewater			Quality of effluent		
		BOD	T-N	T-P	BOD	T-N	T-P
Ham and bacon	Separation of oil Submerged filter Coagulation-sedimentation	2800	160	92	5	13	23
Lard for eating	Separation of oil Activated sludge	2810	320	64	3	10	1.6
Milk and dairy cream	Activated sludge	2530	49	5	2	1	0.5
Flavor enhancer	Activated sludge	2040	1240	51	4	72	1
Vegetable oils and spreads	Activated sludge	1600	27	150	56	8	24

Table 2. Examples of wastewater treatment performance in food manufacturing industries.

2.3.2. Chemical industry

Pharmaceuticals	BOD 40-2500, T-N 80-100, T-P 10-20
Dye works	BOD 10-350, T-N 25, T-P 10
Pulp products	BOD 100-150
Petrochemicals, coal products	BOD 20-200, T-N 20-30, T-P 5
Laundries	BOD 90-1450, T-N 10-40, T-P 10-40
Iron manufacturing	BOD 3000-4000, T-N 800-1000, T-P 20-50,cyanide, phenol
Non-ferrous metal manufacturing	pH 2-8, SS 70-3000, heavy metals

Table 3. Water quality of wastewater from chemical industries

Kind of industry	Products	Sources of wasteater	Quality of wastewater and hydraulic loading	
Leather	Tanned	Raw material	pH	7-12

manufacturing factories	leather, chrome tanned leather, aquatic leather and coloring leather	treating facility, tanning facility, lime immersion facility, dyehouse and chrome immersion facility	BOD CODMn SS Cr T-N T-P hydraulic loading	80-2500 mg/l 100-2000 mg/l 50-3000 mg/l 50 mg/l 250-350 mg/l 10-20 mg/l 30-600 m^3/day
Artificial fertilizer manufacturing factories	Ammonite, nitrolime, phosphate fertilizer and complete fertilizer	Reaction facility, facility for washing exhaust	pH BOD CODMn SS T-N T-P hydraulic loading	1-4 800-1200 mg/l 1000-1500 mg/l 50-350 mg/l 250-350 mg/l 220-280 mg/l 60-100 m^3/day
Inorganic products manufacturing factories	Products of alkali manufacture, calcium, carbide, artificial graphite and phosphate	Reaction facility, washing facility and facility for washing exhaust	pH BOD CODMn SS T-N T-P hydraulic loading	1-9 20 mg/l 40 mg/l 1000-2000 mg/l 60-100 mg/l 2-50 mg/l 500-2000 m^3/day
Electroplating factories	Electroplating	Electroplating facility, alkali cleaning facility and degreasing facility	pH CN Cr^{6+} Zn containing Cu, Cd, Fe, Ni trichloroethylene hydraulic loading	2-12 2-190 mg/l 0.3-590 mg/l 1-830 mg/l 2-90 mg/l 50-2000 m^3/day

Table 4. Sources and water quality of industrial wastewater

The quality of wastewater from the chemical industry is shown in Table 3 and Table 4. Wastewater is discharged from several processes in each industry. Qualities of wastewaters are very different between the industries. The target BOD:N:P ratio for wastewater receiving biological treatment is considered to be 100:5:1 and application of biological treatment is difficult for wastewater if the ratio is very different from this. If nitrogen and phosphorus are low compared to BOD in wastewater, nitrogen and phosphorus should be added in sufficient quantity to permit unrestricted growth of microorganisms. Wastewater from leather manufacturing factories and electroplating factories contain heavy metals and organochloride compounds. In particular, wastewater from electroplating factories is highly toxic when it contains hexavalent chromium and trichloroethylene. Wastewater from non-ferrous metal and iron manufacturing industries are toxic due to the content of heavy metals, cyanide and phenol. Nitrogen concentration is high in the case of leather manufacturing factories, artificial fertilizer factories and the iron industry. Phosphorus concentration is high in effluent from artificial fertilizer factories. Depending on the quality of the wastewater, feasible treatment technologies may include biological, chemical and physical treatments.

Thus the concentrations of organic matter, nitrogen, and phosphorus in industrial wastewater varies greatly according to the type of industry. Organic wastewater is generally treated by biological processes such as activated sludge or the biofilm process. Organic wastewater that contains hazardous substances is first treated by the coagulation-sedimentation process or neutralization method, etc. to remove the hazardous substances; it is then treated by a biological process using activated sludge. Inorganic wastewater containing hazardous substances is treated by a chemical process.

2.3.3. Livestock industry and fish farm

Discharge of untreated livestock excrement from feedlots, and the untreated water used to clean these feedlots, is an extremely serious pollution source. Pollutants discharged from feedlots include organic matter, nitrogen, and phosphorus. The pollution loadings per domestic animals are shown in Table 3 in *Pollution Sources*. The volume of excrement discharged per day is 5.4 kg per pig and 50 kg per cow: both far more than the 1.5 kg per adult human being. The high pollutant loading of the wastewater from feedlots is notorious, but it can be biologically treated. It has a high concentration of nitrogen, and a strong smell. It is, of course, possible to sharply reduce the concentration of pollutants in wastewater from feedlots by removing the feces from the feedlot in advance, so as not to mix feces and water. Wastewater from feedlot is treated by the activated sludge process and anaerobic treatment processes.

Some water bodies bear a heavy pollutant loading from fish farms. There are two types of fish farm, net cage culture located in seas and lakes and fish-farming pond. The pollutant loading produced by keeping fish in fish tanks derives from the organic matter, nitrogen, and phosphorus in the fish feed. In Lake Kasumigaura, the pollution loading caused by carp breeding accounts for 7% of the COD_{Mn} loading, 6% of the nitrogen loading, and 19% of the phosphorus loading. Clearly the phosphorus loading is very high.

3. Countermeasures for point source pollution

The most important point source countermeasure is, of course, treatment of the wastewater at source, such as provision of sewerage, construction of small-scale wastewater treatment plants in regions where it is difficult to construct sewerage treatment systems, and construction of industrial wastewater treatment facilities. In countries around the world, environmental standards and wastewater standards are set as point source loading measures, and effluent standards in Japan are as shown in Table 5 in *Pollution Sources*. The Table shows that the standards for nitrogen and phosphorus that are causes of eutrophication are 120mg/l and 16mg/l respectively, values much higher than the actual nitrogen and phosphorus concentrations in water bodies. In shallow lake and coastal sea areas beside densely populated area, inflow of effluent containing nitrogen at 120mg/l and phosphorus at 16mg/l might exceed the effect of dilution and the self-purification capacity of the water body, thus increasing nitrogen and phosphorus concentration.

	Volume of effluent (m³ day⁻¹)	Newly constructed		Old	
		N	P	N	P
Food manufacturing industry	20≦ V ≦50	20	2	25	4
	50≦ V ≦500	15	1.5	20	3
	500 < V	10	1	15	2
Metal products manufacturing industry	20≦ V ≦50	20	2	30	3
	50≦ V ≦500	15	1	20	2
	500 < V	10	0.5	15	1
Other manufacturing industry	20≦ V ≦50	12	1	15	1.5
	50≦ V ≦500	10	0.5	12	1.2
	500 < V	8	0.5	10	1
Livestock industry	20≦ V ≦50	25	3	50	5
	50≦ V ≦500	15	2	40	5
	500 < V	10	1	30	3
Sewerage	20≦ V ≦100,000	20	1	20	1
	100,000 < V	15	0.5	15	0.5
Sewage disposal	20≦ V	10	1	20	2
Septic tank	20≦ V	15	2	20	4
Other industry	20≦ V ≦50	20	3	30	4
	50≦ V ≦500	15	2	25	4
	500 < V	10	1	20	3

Table 5. Prefectural stringent effluent standards based on the bylaw on eutrophication control of Lake Kasumigaura in Japan

In cases like this where it is difficult to satisfy environmental standards based on uniform national standards, prefectures can enact regulations establishing prefectural stringent effluent standards. Table 5 shows add-on standards for nitrogen and phosphorus established under regulations to prevent eutrophication of Lake Kasumigaura. Standard values for nitrogen and phosphorus vary according to the industry, volume of effluent, and whether it is newly constructed or old. The add-on standards for nitrogen and phosphorus at large-scale final treatment plants of sewage systems are 15 mg/1 and 0.5 mg/1 respectively: values far lower than the national standards of 120 mg/1 and 16 mg/1 respectively.

Because kitchen wastewater accounts for a high percentage of the pollutants in domestic wastewater and contains large quantities of organic matter and suspended material, measures to deal with domestic wastewater, particularly kitchen wastewater, must be implemented in order to conserve the aquatic environment. In consideration of this fact, in 1990, the Water Pollution Control Law was revised by the addition of provisions to strengthen control of domestic wastewater and the clarification of responsibilities of administrative bodies and citizens in implementing domestic wastewater measures. Because domestic wastewater is closely related to the daily lives of people, it refers to measures that should be taken in the kitchens of every home. These include absorbing oils in paper etc., or hardening it with a market coagulant, then throwing it out with the garbage, using appropriate quantities of cleansers. It is claimed that the implementation of kitchen measures, which are citizen participation environmental conservation methods, could cut the pollution loading by between 30% and 50%.

In Japan, local administrative bodies are responsible for designating domestic wastewater measure priority districts, and for enacting domestic wastewater measure promotion plans. Domestic wastewater measure priority districts are designated by prefectural governors as districts where it is considered particularly necessary to promote domestic wastewater measures. Municipalities that are part of domestic wastewater measure priority districts must prepare domestic wastewater measure promotion plans.

Concentration regulations by themselves are inadequate to improve water quality in many areas of enclosed water bodies; the total quantity of pollutants must be reduced. To do this, comprehensive water quality conservation measures that deal with all industrial and domestic wastewater and also with internal loading and non-point source loading must be implemented. In Japan, regulations governing Johkasous installed in individual homes are becoming increasingly strict. For example, in the basin of Lake Kasumigaura in Ibaraki Prefecture, detached dwellings are covered by the standards: BOD 10mg/1 or less, T-N 10 mg/1 or less, and T-P 1 mg/1 or less. From 2004, the Ministry of Environment in Japan introduced a subsidy for advanced JOHKASOU

4. Wastewater treatment processes

4.1. Activated sludge process and modified process

Typical biological wastewater treatment processes are the activated sludge process, the biofilm process, the oxidation pond and the methane fermentation process. In the standard activated sludge process (Figure 5), bacteria and fungi assimilate organic matter and proliferate under aerobic conditions. Protozoa and metazoa predate bacteria and fungi and proliferate. These microorganisms flocculate and form 'activated sludge'. Activated sludge flows into the settling tank and flocculated microorganisms settle down so that a clear supernatant is obtained. The supernatant is discharged into surface water bodies after disinfection. A proportion of the settled microorganisms is returned to the aeration tank to maintain the density of microorganisms. The remainder is concentrated, digested and dehydrated, followed by incineration.

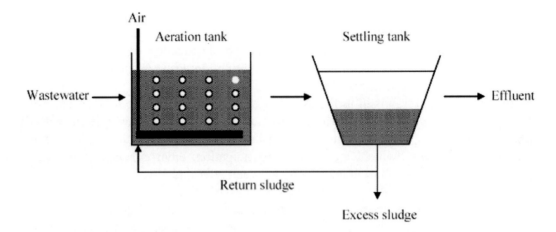

Figure 5. Schematic diagram of standard activated sludge process

The standard activated sludge process has high treatment efficiency of organic matter, but nitrogen and phosphorus removal efficiency is poor. In regions with stringent regulation of effluent water quality for nitrogen and phosphorus, wastewater treatment by the activated sludge process cannot attain the standard. An additional nitrogen and phosphorus removal process must be introduced in regions where the discharge from wastewater treatment facilities might cause eutrophication of aquatic ecosystems. Most nitrogen in domestic wastewater is ammonium nitrogen, and protein in wastewater is decomposed by microorganisms to produce ammonium nitrogen. For nitrogen removal, therefore, both nitrifying and denitrifying bacteria are added in wastewater treatment facilities. For phosphorus removal, phosphorus-accumulating bacteria are used, as well as physical and chemical treatment, e.g. coagulants such as poly-aluminum chloride. Wastewater treatment processes with high nitrogen and phosphorus removal efficiency are shown in Figures 6 and 7.

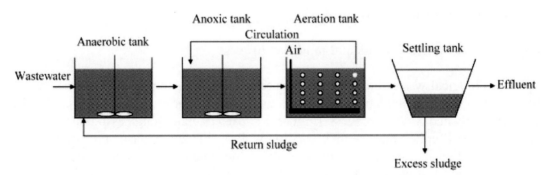

Figure 6. Schematic diagram of A_2O (anaerobic-anoxic-oxic) process

Nitrification occurs in an aeration tank, and denitrification occurs in an anoxic tank. Activated sludge containing nitrate and nitrite by nitrification is circulated to the anoxic tank. Nitrate and nitrite are deoxidized by facultative anaerobic bacteria, to produce gaseous nitrogen. A volume of circulating activated sludge is mixed with 2 to 4 times the volume of influent wastewater. Phosphorus is removed biologically in the A_2O process. Phosphorus accumulating bacteria release phosphate and grow in an anaerobic tank. In the aeration tank, phosphorus-accumulating bacteria assimilate phosphate, increasing the phosphorus content of activated sludge from 2 to 4-8%. By increasing the phosphorus content of activated sludge, high removal efficiency of phosphorus is achieved in A$_2$O process. Coagulants such as poly-aluminum chloride are used for removal of phosphate in anoxic-oxic process with coagulant addition. The compounds produced by reaction of phosphate and coagulants settles down in the settling tank and phosphorus removal is achieved. If this material does not contain hazardous compounds, it can be used as fertilizer, but most of it is disposed in landfills.

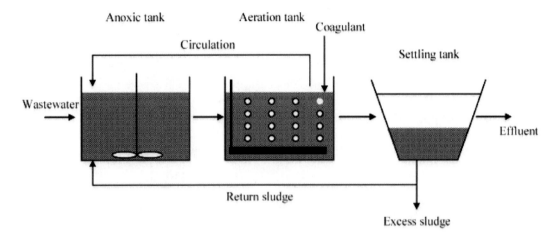

Figure 7. Schematic diagram of anoxic-oxic process with coagulant addition.

4.2. Biofilm process

On the surface of stones in rivers, bacteria, algae, protozoa and metazoa exist and contribute to self-purification of the water. The biofilm process is based on this process of self-purification, reproducing the process artificially and efficiently in a water treatment plant. In the biofilm process, carriers such as plastic, stone, and ceramics are added to the tank to provide a substrate for adhesion of microorganisms. The tank is aerated from the bottom and microorganisms purify the wastewater under aerobic conditions. The different forms of biofilm processes include trickling filter, rotating disc, submerged filter and biological filtration.

In the trickling filter, stones are laid to a depth of 2-3 m and wastewater is sprayed over the stones. A biofilm forms on the surface of the stones and purification proceeds. In the rotating disc process, part of the disc is immersed in a tank filled with wastewater and the disc is rotated like a waterwheel. A biofilm forms on the surface of the disc. Oxygen is distributed to microorganisms when the biofilm is in contacts with air. In the submerged filter system, a plastic carrier is fixed in the aeration tank. A biofilm develops in the aerated flowing wastewater. In the biological filtration process, a granular carrier such as sand, activated carbon, plastic and ceramic is loaded into the aeration tank. A biofilm forms on the surface of the carrier and purification proceeds by biofilm. The surface area per volume is extremely high because the size of the carrier is small and the material is porous. This feature enables development of a compact Johkasou.

In the aeration tank of the activated sludge process, organisms which proliferate slowly are washed out, but such organisms can survive in the biofilm process attached to the carrier. The diversity of organisms in the biofilm process is higher than that of activated sludge. Macroorganisms which do not appear in activated sludge occur in the biofilm process, making the food chain longer and more complex. Sludge production is lower in the biofilm process than in activated sludge because of the long food chain.

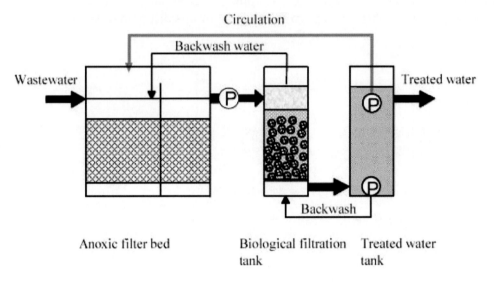

Figure 8. Schematic diagram of anoxic filter bed- biological filtration process

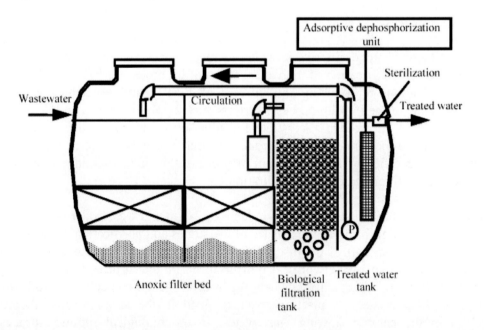

Figure 9. Schematic diagram of an anoxic filter bed - biological filtration process using ceramics made from sediment of eutrophic lakes and an adsorptive dephosphorization unit

For removal of nitrogen, an anoxic tank is introduced to the biofilm process. An anoxic filter bed-biological filtration process is shown in Figure 8. A plastic and a ceramic carrier are installed in the anoxic filter bed and the aerobic biological filter respectively.

For denitrification, water treated in the water tank is returned to the anoxic filter bed. In a biological filtration tank, a backwash is executed once a day for regeneration of the biofilm and maintaining the correct thickness of biofilm. Exfoliated biofilm is sent to the first anoxic filter bed. The activity of microorganism is thus maintained. When domestic wastewater is treated by this process, the quality of the treated water is: BOD < 10 mg/l, total nitrogen < 10 mg/l and suspended solids < 10 mg/l. For utilization of unused material and improvement of phosphorus removal, an anoxic filter bed - biological filtration process using ceramics made from sediment of eutrophic lake and adsorptive dephosphorization unit has been developed (Figure 9). As When domestic wastewater is treated by this process, the quality of the treated water is: BOD $\leq$10mg/l, total nitrogen $\leq$10mg/l and total phosphorus $\leq$1mg/l.

4.3. Anaerobic treatment

In anaerobic treatment, organic matter in industrial wastewater and excess sludge is decomposed to carbon dioxide and methane by anaerobic bacteria. This anaerobic treatment is called the methane fermentation process. Anaerobic treatment has advantages such as low energy consumption, as there is no need of aeration, and the methane can be used as a fuel. This process is suitable for high strength wastewater, providing destruction of pathogenic microorganisms and helminth eggs. In anaerobic treatment, organic matter is decomposed in three steps—hydrolysis, acid production and methane production (see Figure 10). Facultative anaerobic bacteria and obligate anaerobic bacteria contribute to purification of wastewater.

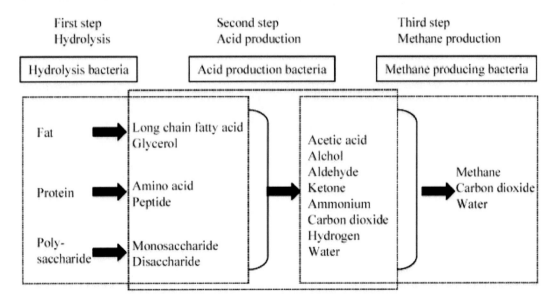

Figure 10. Decomposition process of organic matter by anaerobic treatment

Anaerobic treatment is executed after pretreatment such as removal of earth, sand and impurities, pH adjustment, control of wastewater concentration and flow rate. This treatment produces gas, treated water and sludge. The gas is utilized for generating electricity, after removal of hydrogen sulfide. The treated water is then treated under aerobic condition for removal of organic matter. Sludge is incinerated after dehydration. Anaerobic treatment is classified into mesophilic fermentation and thermophilic fermentation. These are operated at 36-38 °C and 53-55 °C respectively. Mesophilic fermentation is generally more common. The UASB (upflow anaerobic sludge blanket) process is one example of the anaerobic treatment process (see Figure 11). UASB is used for treatment of high strength wastewater such as wastewater from breweries, because it can maintain a high concentration of microorganism and high treatment efficiency. Wastewater flows from the bottom to the upper part of an anaerobic tank, enabling microorganism to naturally form granules (a floc of microorganisms). In the UASB process, the suspended mass of microorganisms can be as high as 50 000 mg/l, approximately ten times higher than with activated sludge and other anaerobic treatment. Consequently, UASB can be operated with a ten times higher loading of organic matter than is possible with activated sludge.

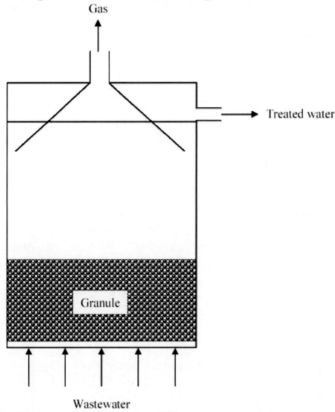

Figure 11. Schematic diagram of UASB process

Thus UASB offers high removal efficiency of organic matter, but it has the demerit of low nitrogen and phosphorus removal efficiency and inhibition of granule formation if the wastewater has a high content of suspended solids.

By combination of the UASB and aerobic biofilm processes, wastewater from breweries with BOD of more than 1000 mg/l can be successfully treated. The BOD reduction rate can be 98% and the nitrogen removal rate more than 70%, for wastewater with loadings of 10 kg-COD_{Mn}/m^3/day and 0.4 kg-T-N/m^3/day.

4.4. Thermophilic oxic process

Thermophilic oxic process is suitable for wastewater with high concentration of suspended solids. In this process, production of surplus sludge is very low. Organic matter is decomposed by bacteria at 60 °C. The principle of the thermophilic oxic process is similar to the composting process, but the thermophilic oxic tank is filled with wood chip to 80% of volume as a promoter of fermentation, and aeration is executed for improvement of removal efficiency. In this process, shochu (distilled liquor) wastewater with BOD 70 000 mg/l can be treated efficiently under the following operational conditions: BOD loading 3.5 kg/m^3/day, hydraulic loading 50 L/m^3/day and aeration 100 L/m^3/min. The thermophilic oxic process is used for treatment of surplus sludge from the activated sludge process, garbage, waste from livestock farming and high strength organic wastewater from food manufacturing factories.

Glossary

A$_2$O process:	Modification process of standard activated sludge process for enhanced nitrogen and phosphorus removal, consisting of anaerobic tank, anoxic tank, aeration tank and sedimentation tank. In the anaerobic tank, phosphorus-accumulating bacteria proliferate. In the anoxic tank, denitrification by facultative anaerobic bacteria proceeds. In the aeration tank, nitrification by nitrifying bacteria and phosphate absorption by phosphorus accumulating bacteria proceed. A proportion of the activated sludge forming in aeration tank is returned to the anoxic tank for denitrification.
Activated sludge process:	Typical wastewater treatment process consisting of aeration tank and sedimentation tank. In aeration tank, microorganisms such as bacteria and fungi assimilate organic matter and proliferate at aerobic condition, and protozoa and metazoa predate bacteria and fungi, forming 'activated sludge'. The activated sludge next flows into a sedimentation tank, followed by settling of microorganisms, and clear water is obtained. The clear water is discharged to water environment after disinfection. Settled microorganisms are returned to the aeration tank or are incinerated after digestion treatment and dehydration.
Advanced Johkasou:	JOHKASOU (JOHKA means purification and SOU means tank) have been developed as on-site domestic wastewater treatment systems. Old type JOHKASOU were able to remove only BOD. Advanced JOHKASOU have functions which can improve effluent quality with regard to BOD, nitrogen and phosphorus, to less than 10 mg/l, 10 mg/l and 1 mg/l respectively. This type has been recognized as important in control of eutrophication in enclosed water bodies such as lakes and marshes, and inland seas.
Algal bloom:	The phenomenon of algae accumulating in lakes, reservoir and seas. The water may be colored green, red or brown.
Anaerobic tank:	Part of A$_2$O process. Dissolved oxygen and combinative oxygen (NO$_3^-$, NO$_2^-$) are inexistent.
Anoxic tank:	Part of A$_2$O process. Dissolved oxygen is inexistent, but combinative oxygen (NO$_3^-$, NO$_2^-$) is present.
Biofilm process:	A wastewater treatment process using microorganisms attached to a carrier. Carriers such as plastic, stone, and ceramic are added to the

	tank to provide a substrate for adhesion of microorganisms.
Biological filtration process:	One of the biofilm processes. A granular carrier such as sand, activated carbon, plastic and ceramic is added to the aeration tank. A biofilm forms on the surface of the carrier and purification proceeds by biofilm. The surface area per volume is very high because the size of the carrier is small.
BOD (Biochemical Oxygen Demand):	An index of pollution by organic matter, measuring consumption of oxygen by respiration of aerobic microorganisms assimilating organic matter and growing in water at constant condition.
Coagulation sedimentation process:	A wastewater treatment processes for removal of suspended solids and heavy metals. By addition of coagulant, suspended solids and metal hydroxide coagulate and sedimentation rate increase. Thus suspended solids and heavy metals precipitate and treated water is obtained. As coagulant, aluminum sulfate, aluminum chloride, ferric chloride and macromolecule coagulants are used.
COD$_{Mn}$:	Chemical Oxygen Demand using potassium permanganate as an oxidizer. An index of pollution by organic matter.
Eutrophication:	Elevated levels of nitrogen and phosphorus (originated from human activities) in lakes, reservoirs, inland sea, increase the rate of primary production, often leading to algae blooms and red tides.
Facultative anaerobic bacteria:	Bacteria growing under anaerobic and aerobic conditions.
Metazoa:	All kind of animals excluding protozoa. The body consists of many cells, and their morphology and function are specialised.
Neutralization:	Neutralization is one of the treatment processes for wastewater containing heavy metals. Wastewater containing metal ions is normally acidic. Addition of alkali increases pH and decreases the solubility of the metal ions. Consequently metal hydroxide precipitates and treated water is obtained.
Obligate anaerobic bacteria:	Bacteria only growing under anaerobic conditions.
Pathogen:	Microorganism causing diseases of human and animal. Specific protozoa, bacteria and virus are known as pathogens.
Protozoa:	Unicellular lower animals inhabiting water and soil. Some species are parasitic.
Rotating disc:	One of the biofilm processes. Part of a disc is immersed in a tank filled with wastewater and the disc is rotated like a waterwheel. A biofilm forms on the surface of the disc. Oxygen is distributed to microorganisms when the biofilm is in contact with air.
Sewage works:	A facility for removing suspended solids, organic matter, nitrogen and phosphorus from wastewater and discharging treated water to rivers, lakes and seas.
Submerged filter:	One of the biofilm processes. A plastic carrier is fixed in an aeration tank. A biofilm grows in the flowing aerated wastewater.
T-N:	Total Nitrogen.
T-P:	Total Phosphorus.
Trickling filter:	One of the biofilm processes. As a carrier, stones are laid to a depth of 2-3 m and wastewater is sprayed over the stones. A biofilm forms

on the surface and purification proceeds as the wastewater slowly trickles down.

Bibliography

Japan Environmental Management Association for Industry editorial supervision (1995). Industrial pollution control technologies and laws- water quality- [in Japanese]. Maruzen.

Yasuhiro Yamamoto *et. al.* (2002). Evaluation of anaerobic filter bed- biological filtration process with physicochemical phosphorus removal methods, Japanese Journal of Water Treatment Biology, 38, 47-55 (in Japanese).

Biographical Sketches

Yuhei Inamori is an executive researcher at the National Institute for Environmental Studies (NIES), where he has been in his present post since 1990. He received B.S. and M.S. degrees from Kagoshima University, Kagoshima Japan, in 1971 and 1973 respectively. He received a PhD degree from Tohoku University, Miyagi, Japan in 1979.

From 1973 to 1979, he was a researcher for Meidensha Corp. From 1980 to 1984, he was a researcher for NIES, and from 1985 to 1990, a senior researcher for NIES. His fields of specification are microbiology, biotechnology and ecological engineering. His current research interests include restoration of aquatic environments using Bio-Eco engineering applicable to developing countries. He received an Award of Excellent Paper, from the Japan Sewage Works Association in 1987, an Award of Excellent Paper from the Journal "MIZU" in 1991, an Award of Excellent Paper, from the Japanese Society of Water Treatment Biology in 1998, and an Award for Environmental Preservation from the Prime Minister of Korea in 1999.

Naoshi Fujimoto is Assistant Professor of Tokyo University of Agriculture, where he has been at his present post since 2000. He received B.E. and M.E. degrees from Tohoku University, Miyagi, Japan, in 1991 and 1993 respectively. He received a D.E. degree in civil engineering from Tohoku University in 1996. From 1996 to 1999, he was a research associate of Tokyo University of Agriculture. His specialized field is environmental engineering. His current research interests include control of cyanobacteria and its toxins in lakes and reservoirs.

NON-POINT SOURCES OF POLLUTION

Yuhei Inamori
Executive researcher, National Institute for Environmental Studies, Tsukuba, Japan

Naoshi Fujimoto
Assistant Professor, Faculty of Applied Bioscience, Tokyo University of Agriculture, Tokyo, Japan

Keywords: organic matter, nitrogen, phosphorus, rainfall, patty field, arable land, fertilizer, forest, urban runoff

Contents

1. Introduction
2. Definition of a Non-point Source
3. Non-point Sources and its Loads
4. Countermeasures for Non-point Source Pollution
Glossary
Bibliography
Biographical Sketches

Summary

Non-point source pollution loads are loads with their discharge sources spread over wide areas. The precise locations of non-point source loads cannot be identified as they are dispersed throughout the catchment area. It is generally difficult to regulate non-point sources because their discharge locations cannot be clarified. Forms of land use that cause non-point source loads include farmland, forests and urban areas.

The non-point source loads as a percentage of the total load in Lake Kasumigaura in Japan are 44% for organic matter (COD), 42% for nitrogen, and 23% for phosphorus.

Approximately 30% of fertilizer used in arable land becomes non-point source pollution. In paddy fields, discharge of water containing mud and fertilizer is the main pollution source.

Livestock excrement affects water quality of river and groundwater as non-point source where livestock are at pasture, breeding density is high and management of excrement is unsatisfactory.

When there is an increase of paved area, there is an increase of rainwater discharge to water bodies because penetration of rainwater to the ground is inhibited. Pollutants accumulated in urban area are washed out by rainfall and flow into water bodies.

In agriculture, slow release fertilizers coated with material to control the elution of nutrients is used as a countermeasure against non-point source pollution. Moreover, modified fertilization methods which fertilize around the rice seeding and control the diffusion of fertilizer into the water, are becoming more popular.

To control discharge from urban areas, ditches and residential land should be cleaned with the cooperation of residents through information and educational programs—reducing the quantity of pollutants discharged by rainwater. Control of surface flow by flood storage basins, and permeable pavements to facilitate infiltration, are useful measures to reduce pollution from urban areas.

It is important to retain areas of forest for conservation of water bodies as they absorb nutrients and other pollutants.

1. Introduction

Non-point source pollution loads are loads with their sources dispersed over wide areas. These loads flow into water bodies including rivers, lakes, reservoirs, ponds, wetlands, groundwater, and the sea, and they may contain organic matter, nitrogen, phosphorus, and agricultural chemicals, causing pollution and eutrophication. Rainfall, discharge from agricultural fields, cleared forest and urban runoff are typical non-point sources. Measures to effectively deal with these sources are extremely important because the contribution of non-point source to total pollution is increasing as a result of improved wastewater treatment facilities. In this chapter, a definition is given for non-point source pollution; amounts of non-point source loads are given and countermeasures for non-point source pollution are described.

2. Definition of a Non-point Source

Non-point source loads are those whose discharge sources are dispersed over wide areas: the discharge of fertilizer and agricultural chemicals from agricultural land, rainfall containing atmospheric pollutants, exhaust gases, discharge of sediment accumulated on roofs, roads, and the ground surface, animal excreta, carcases of dead animals, fallen leaves in urban area, and discharge of pollutants from clear-felled parts of forests. The amount of precipitation is closely related to non-point source pollution loads because rainwater conveys pollutants to water bodies.

The precise locations of non-point source loads cannot be identified as they are scattered throughout the catchment area. Non-point sources flow through channels, storm sewers and groundwater and eventually into rivers, lakes, reservoirs and the ocean. It is generally difficult to regulate them because their discharge locations cannot be clarified. They may contain organic matter, nitrogen, phosphorus and agricultural chemicals. Non-point sources cause eutrophication and disruption of ecological balance. Measures to effectively deal with these sources are therefore extremely important.

The rise in the concentration of agricultural chemicals in river waters when agricultural chemicals are spread on fields is regarded as non-point source pollution. Forests play a role in purifying rainfall because the nitrogen and phosphorus concentrations in surface water draining from forests are lower than those of the rainfall. If the forest is cleared, soil and organic matter are washed out by rainfall leading to organic pollution and eutrophication. Even in forest without human activity, organic matter, nitrogen and phosphorus loads in the drainage water may be high in autumn and winter, as the dead plant material decomposes. In spring and summer, however, the nutrients are absorbed by the growing vegetation.

It is generally not possible to take localized measures to deal effectively with non-point source pollution loads because they exist over wide areas. More precise inventories reveal that in many regions their contribution to total pollution is higher than had been previously estimated.

The non-point source loads as a percentage of the total load in Lake Kasumigaura in Japan are 44% for organic matter (COD), 42% for nitrogen, and 23% for phosphorus (see Figure 1 in *Point Sources of Pollution*). As it is important to clarify the source and quantity of inflowing pollutants in order to establish water pollution prevention measures, the load per unit of area has been calculated for each non-point source load: organic matter, nitrogen, and phosphorus (see Table 1). A comparison of pollutant loads from paddy fields, arable land, forests, golf courses and cities, shows that the load from cities is relatively high. It is, therefore, predicted that the urbanization of a catchment area increases its non-point source loads.

Pollutant	Paddy field	Arable fields	Mountains and forests	Golf course	Rain	Urban runoff
COD	7.19	2.45	3.83	3.83	6.95	15.3
Nitrogen	2.4	2.34	1.56	1.56	3.08	2.4
Phosphorus	0.095	0.116	0.054	0.054	0.13	0.18

unit: $kg \cdot km^{-2} \cdot day^{-1}$, averaged over 1 year.

Table 1. Amount of pollution source used for estimation of non-point source pollution of Lake Kasumigaura

3. Non-point Sources and its Loads

3.1. Rainfall

The pollution load being received by a water body and its catchment area from the atmosphere is included in non-point sources. Nitrogen and phosphorus loads from air reach the land surface as dry fallout and wet fallout—rain and snow. The concentration of nitrogen in rainwater is reported to range from 0.5 to 1.2 mg/l as T-N. The nitrogen load from rain is large because the concentration is as high as that of a eutrophic lake. The proportion of dry fallout to total fallout accounts for approximately 40%. Nitrogen oxide in rainwater is emitted from factories and automobiles. Nitrogen concentration in rain is high in urban and industrial areas, and low in mountains and isolated islands. Phosphorus concentration in rain is reported to range from 0.005 to 0.1 mg/l.

3.2. Forest

The pollution load from forest per area is lower than other non-point sources such as paddy fields and urban runoff (Table 1). Pollution loads from forests and mountains vary widely. Minimum pollution loads from forests and mountains in Japan are reported to be 0.085 kg/km^2/day for nitrogen, 0.027 kg/km^2/day for phosphorus and 1.68 kg/km^2/day for COD. These loads are the lowest of the pollution loads from any other land use. Thus conservation of areas of mountain and forests without fertilization is important for reduction of pollution load.

In regions where natural forests occupy vast areas, however, such as much of Japan, the total pollution load from forests comprises a large proportion of the total pollution load. Rainwater falling on a forest is in contact with trees and soil and is affected chemically and biologically. Some of the water infiltrates into the ground and subsequently discharges slowly, and some of it is discharged more quickly as surface runoff. The pollutants originating from forest are consequently litter, soil and dissolved matter. The quality of water discharging from a forest fluctuates during a rainfall episode. Particulate nitrogen and phosphorus concentration increase steeply during rainfall as the concentration of particulate matter increases in the runoff. Concentrations of nitrate and phosphate increase with increase of flow rate from forest because the nitrate and phosphate accumulated from biological decomposition flows out in the surface runoff. Nitrogen and phosphorus concentrations in water discharged from forests are lower than those of rainfall demonstrating the purification role played by forests. The quality of water discharging from forests depends on geology, soil, the species of tree and their age, temperature, etc.

3.3. Agriculture

In Japan, the water used in agriculture accounts for 64.1% of total water use. Approximately 95% of the water used in agriculture goes to paddy fields. Clearly, therefore, the pollution loads from paddy fields is very important. In the rice planting season, in order to accommodate rice planting machines, water containing mud and nutrients is discharged to lower the water level. At this time, the nitrogen and phosphorus loads comprise 24% and 22% of the total load in the irrigation period. The application of fertilizer in paddy fields during the summer is accompanied by a rise in the nitrate nitrogen concentration in groundwater near the paddy fields. It has been reported that the nitrate nitrogen of groundwater near paddy fields in January (off-season) is 2 mg l^{-1} or less, but in June (the growing season), it rises to a maximum of 13 mg l^{-1}. It is known that if infants drink groundwater contaminated by a high concentration of nitrate, they may develop methaemalbuminemia, a condition with symptoms of cyanosis and that can produce nitrosoamines which are carcinogens. In arable land, nitrogen loads increase with increased fertilization application.

Pollution loads from paddy fields are estimated to be 2.3 to 6.9 kg/km^2/day for nitrogen, 0.09 to 0.76 kg/km^2/day for phosphorus and 5.2 to 23.3 kg/km^2/day for COD in Japan. Pollution loads from arable land are estimated at 2.2 to 14.4 kg/km^2/day for nitrogen, 0.05 to 0.17 kg/km^2/day for phosphorus and 1.3 to 26.0 kg/km^2/day for COD in Japan. Approximately 30% of fertilization is lost and becomes a non-point source pollution load.

In paddy fields, an anaerobic layer forms and there is a certain amount of denitrification of the nitrate nitrogen. Because an anaerobic layer does not form on fertilized arable lands, denitrification does not normally occur.

Livestock wastes such as excrement are classed as point sources if they are purified in a wastewater treatment plant, but their reuse as fertilizer after composting or fermentation, leads to a non-point pollution source. Reuse of livestock wastes in arable land causes increases in nitrogen and phosphorus concentrations in nearby water bodies.

In areas where piggery waste is stored and in livestock feedlots, it was reported that nitrogen concentrations in downstream rivers correlated with livestock breeding density. Nitrate nitrogen in rivers in such areas can be 20 to 40 mg/l with breeding density of 2000 to 4000 head/km^2. Nitrogen permeates into the ground from pits holding livestock waste and is eventually discharged to river.

Table 2 shows an example of the calculation of loads when nitrogen fertilizer, feedlot waste material, and domestic wastewater permeate into the ground.

	Fertilization kg ha^{-1} year^{-1}	Slurry etc. from livestock industry		House discharging effluent through night soil treatment system into ground
		Pig	Cattle	
Amount of nitrogen discharged	135	37 g head^{-1} day^{-1}	290 g head^{-1} day^{-1}	6 g person^{-1} day^{-1}
Load to groundwater	34	3.4 kg head^{-1} year^{-1}	26 kg head^{-1} year^{-1}	1.6 kg person^{-1} year^{-1}

- Amount of nitrogen fertilization was calculated as average amount of Japanese field.
- Nitrogen loads into ground water were calculated by 25% leaching into groundwater for fertilization and livestock.
- Nitrogen load was calculated for livestock as no nitrogen in sewage was removed.
- Nitrogen load from house discharging effluent through night soil treatment system into ground was based on an experiment of soil trench system with lysimeter treating 50 L m^{-2} day^{-1} effluent from night soil treatment.

Table 2. Nitrogen sources and loads in ground water

3.4. Urban runoff

In urban areas, accumulated pollutants increase with higher levels of production activity and physical distribution. As more and more of the ground area is rendered impermeable, the proportion of rainwater discharging to water bodies increases, as penetration into the ground is inhibited. Pollutants accumulated in urban area are washed out by rainfall and flow into water bodies. In urban area, discharge of pollutants by rainwater is an important non-point source. Pollutants accumulated on roads, roofs and vacant land are discharged to sewers or storm sewers. There are two kinds of sewerage system: one mixes sanitary sewage and storm sewage in the same sewer and treats them together, while the other keeps stormwater separate from sanitary sewage and does not treat the stormwater, which instead is discharged to water bodies. In the latter case, pollutants carried by rainwater are discharged to water bodies without treatment. Urban runoff is thus discharged as a non-point source. In the former case, pollutants brought by rainwater are treated in sewage treatment plants. Stormwater and sanitary sewage flows can exceed, by a factor of more than three, the flow of sanitary sewage, causing sanitary and storm sewage to be jointly discharged direct to water bodies. Fallen leaves from roadside trees also create a heavy pollution load when they are discharged by rainwater—obviously, they contain high levels of organic matter, nitrogen and phosphorus. Fertilizer used in urban parks can also find its way into water bodies. Pollution loads of COD and phosphorus from urban areas are estimated to be the highest of the non-point sources (see Table 1).

4. Countermeasures for Non-point Source Pollution

Because it is easier to carry out measures to control point-source loads, the implementation of point-source load measures is accompanied by a relative rise in the non-point source share of total pollution load, thus increasing the importance of non-point source load measures. But non-point source load measures are extremely difficult to implement because pollutants flow into rivers, lakes, reservoirs, etc. from innumerable points.

Lake water quality conservation measures for Lake Kasumigaura in Japan include three non-point source load measures: "promoting environmentally-friendly agriculture", "controlling discharge of loads in urban regions", and "appropriately managing forests". "Promoting environmentally-friendly agriculture" involves informing and educating farmers about the water quality of Lake Kasumigaura and water purification, through public relations, and implementing measures with the cooperation of the farmers. While doing this it is necessary to take care to preserve harmony with productivity, advocating low input agriculture intended to lower environmental loads caused by the use of chemical fertilizers and agricultural chemicals, as well as through appropriate soil preparation and rational crop rotation. Specific important reduction measures are optimizing the quantity of fertilizer applied through soil diagnosis and applying it appropriately by promoting the use of slow release fertilizer. These fertilizers, which have recently been developed, are coated with a material that controls the elution of fertilizer. For example, one product gives 80% elution within 30 to 100 days at 25 °C. A modified fertilization method which fertilizes around rice seedlings and controls diffusion of fertilizer to water has become popular. These methods have resulted in a 20% decrease of fertilizer use without significant loss of production. In paddy fields, the pollution load tends to increase with higher water use. So it is beneficial to reduce water use in order to decrease pollution load.

Another countermeasures against pollution from agriculture, is to temporarily retain water containing fertilizer in fallow fields in low-lying areas, so that it is treated with a natural purification system. A paddy field filled with water serves a nitrogen removal function. If there is arable land above the paddy field, discharged water from the arable land should be conducted into the paddy field and treated biologically.

To control discharge of pollutants from urban areas, ditches and residential land should be cleaned, with the cooperation of residents, through information and educational programs, to reduce the quantity of pollutants discharged from urban regions by rainwater. For reduction of pollution loads from urban areas, the use of storage basins for collecting urban wastewater and rainwater, and enhanced underground infiltration of water by use of permeable pavements, are important. It is also important to provide as much green belt land as possible in newly developed districts. Treatment of storage water by underground infiltration, constructed wetlands and coagulation-sedimentation processes have all been executed for reduction of pollutant from urban areas. Cleaning of roads for removal of sediments is another useful countermeasure.

A pollution load is discharged by rainwater as a result of erosion and collapse of soil in areas of cleared forest. Because properly managed forests make a contribution to reducing rainwater load and creating water resources, the management and expansion of forests in the Lake Kasumigaura drainage basin are extremely important as measures to reduce non-point source loads. Forests reduce the nitrogen and phosphorus loads in rainfall—their concentrations in water draining from forests are lower than those in rainfall. It is very important, therefore, to retain areas of forest for conservation of water bodies. It is also effective to select kind of tree whose discharge of pollution is low.

Glossary

COD: Chemical Oxygen Demand using potassium permanganate as an oxidizer. An indicator of pollution of organic matter.

Coagulation-sedimentation process: A water purification method introduced in waterworks and wastewater treatment facilities for removal of suspended solids. A coagulant such as aluminum sulfate etc. is mixed with raw water so that suspended solids flocculate in a flocculation pond and settle quickly in a settling pond.

Eutrophication: Inflow of nitrogen and phosphorus originating from human activities increase concentration of nutrients in lakes, reservoirs, and inland seas. Primary production increases, often leading to algal blooms and red tide.

T-N: Total Nitrogen.

T-P: Total Phosphorus.

Bibliography

Foundation of River and Watershed Environment Management (2005). Kasen to eiyouenrui. (In Japanese), 179 p., Gihodoshuppan, Tokyo.

Inamori, Y., Fujimoto, N. and Kong, H.N. (1997). Recent aspect and future trend of countermeasures for renovation of nitrate-contaminated groundwater. (In Japanese). Journal of Water and Waste, vol.39, 936-944.

Inoue, T. (2003). Pollutant Load Factor from Non-point Source: Present State and Task for the Future. (In Japanese). Journal of Japan Society on Water Environment, vol. 26, 131-134.

Kunimatsu, T. and Sudo, M. (1997). Measurement and Evaluation of Nutrient Runoff Loads from Mountainous Forested Lands. (In Japanese). Journal of Japan Society on Water Environment, vol. 20, 810-815.

Sudo, R. and Onuma, K. (2000). Water Quality of Lake Kasumigaura and its Non-Point-Source Pollutants. (In Japanese). *Environmental Conservation Engineering*, vol. 29, pp. 509-515.

Takeda, I. (1997). Non-point Source Pollutants in Agricultural Lands. (In Japanese). Journal of Japan Society on Water Environment, vol. 20, 816-820.

Biographical Sketches

Yuhei Inamori is an executive researcher at the National Institute for Environmental Studies (NIES), where he has been in his present post since 1990. He received B.S. and M.S. degrees from Kagoshima University, Kagoshima Japan, in 1971 and 1973 respectively. He received a PhD degree from Tohoku University, Miyagi, Japan in 1979.

From 1973 to 1979, he was a researcher for Meidensha Corp. From 1980 to 1984, he was a researcher for NIES, and from 1985 to 1990, a senior researcher for NIES. His fields of specification are microbiology, biotechnology and ecological engineering. His current research interests include restoration of aquatic environments using bioengineering and eco-engineering applicable to developing countries. He received an Award of Excellent Paper, from Japan Sewage Works Association in 1987, an Award of Excellent Publication from the Journal "MIZU" in 1991, and an Award of Excellent Paper, from the Japanese Society of Water Treatment Biology in 1998.

Naoshi Fujimoto is Assistant Professor of Tokyo University of Agriculture, where he has been at his present post since 2000. He received B.E. and M.E. degrees from Tohoku University, Miyagi, Japan, in 1991 and 1993 respectively. He received a D.E. degree in civil engineering from Tohoku University in 1996. From 1996 to 1999, he was a research associate of Tokyo University of Agriculture. His specialized field is environmental engineering. His current research interests include control of cyanobacteria and its toxins in lakes and reservoirs.

SALINIZATION OF SOILS

Hideyasu Fujiyama
Professor of Agriculture, Tottori University, Tottori, Japan

Yasumoto Magara
Professor of Engineering, Hokkaido University, Sapporo, Japan

Keywords: Capillary effect; irrigation; leaching; osmotic pressure; salinity; salinity hazard; sodium; sodium hazard

Contents

1. Introduction
2. Causes of salinization and saline soil
3. Salinity of water and soil
4. Prevention of salinization and improvement of saline soil
Glossary
Bibliography
Biographical Sketches

Summary

Most irrigation water contains some salts. After irrigation, the water added to the soil is used by the crops or evaporates directly form the moist soil. The salt, however, is left behind the soil. Therefore, unless removed, it accumulates in the soil. This phenomenon is called salinization. Salts in the irrigation water or soil cause an adverse effect on crop production.

To sustain production of crops, irrigation procedures need to be adjusted to control the salts in the soil, as well as controlling the salt concentration in the irrigation water. Electrolytic conductivity, cation exchange capacity, sodium adsorption ratio and other physico-chemical parameters, are used for assessing the salinity of soil and water.

In order to mitigate the effects of salinization to maintain agricultural productivity, several measures have been practiced. The rise of groundwater is mainly caused by the capillary effect of the soil. Irrigation methods to avoid capillary rise have been developed. The most widely used method to improve salinity is leaching. Also, salt-tolerant crops are developed with the help of biotechnology, in order to be able to thrive in a saline environment.

1. Introduction

Normally, soil is rich in salts because the parent rock of the soil contains ionic substances. Seawater is another source of salts in low-lying area along the coast. A very common source of salts in irrigated soils, however, is the irrigation water itself. Most irrigation water contains some salts. After irrigation, the water added to the soil is used by the crops or evaporates directly form the moist soil. The salt, however, is left behind in the soil. Therefore, unless removed, it accumulates in the soil. This phenomenon is called salinization. A white layer of dry salt is sometimes observed on very salty soil. Salty groundwater may also contribute to salinization. When the water table rises, the salty groundwater may reach to the uppermost stratum, thus, to the root zone. Soil which contains excessive amounts of salt is often referred as salty or saline soil. Soil or water which has a high concentration of salt is said to have high salinity.

2. Causes of salinization and saline soil

In many places in the world, the productivity of soil has deteriorated because of an excess of salt has accumulated in the soil around the plant root zone. Large-scale soil salinization has mostly occurred in arid and semi-arid regions. Soil affected by salt also widely exists in sub-humid and humid (i.e. high rainfall) regions. Saline soil is particularly frequent in coastal areas since the soil in those areas is exposed to seawater. Even if the water is low in salinity, the salinity in the soil will increase if the water is used for irrigation for a long time because the trace amount of salt gradually accumulates.

Excessive salinity of the soil surface and the root zone are typical properties of saline soils. The main source of salts in soil is exposed bedrock in geologic strata in the Earth's crust. Salts are gradually released from the bedrock after becoming soluble through physical and chemical weathering such as hydrolysis, hydration, dissolution, oxidation, and carbonation. The released salts dissolve into the surface water or groundwater. As the water with dissolved salts flows from humid regions to less humid or arid regions, salts in the water are gradually concentrated.

The most dominant ions at the place where salts become soluble by weathering are carbonate and bicarbonate of calcium, magnesium, potassium and sodium, if carbon dioxide exists. At first, the salinity of the water is low, but as the water flows from a humid area to a less humid area, it becomes higher as the water evaporates. As the salts in the water are further concentrated, salts with lower solubility start to precipitate. In addition, due to other mechanisms such as ion exchange, adsorption, and the difference of mobility, the concentrations of chemical substances dissolved in the water gradually shift; this always results in increased concentration of chloride and sodium ions in water and soil.

2.1. Classification of saline soil

Saline soil is usually categorized into the following three types, saline, sodic, and alkaline sodic soil.

Saline soil contains a lower amount of sodium absorbed onto soil particles. This type of soil is often seen in sandy soil containing lower amounts of clay and organic matter. Saline soil in deserts is usually of this type.

Sodic soil contains a large amount of sodium absorbed onto soil particles. This type of soil is often seen in soil that contains large amounts of clay.

Alkaline sodic soil is a type of sodic soil that is highly alkaline with the pH value more than 8.5. This type of soil contains higher amounts of carbonate and bicarbonate which can be hydrolyzed to alkalize the products. This soil type has also been called "alkaline soil". Excessive amounts of carbonate and bicarbonate salts may be brought into soil with groundwater by capillary effect or by irrigation water, or may be formed from soil particles themselves.

2.2. Inhibition of plant growth

The growth of a plant is inhibited by salts when the concentration in the soil around the root zone exceeds the plant's salt-tolerance. Excessive salinity in water causes growth inhibition, partial dwarfism, and lower yield, depending on the salinity level. The direct result of excessive salinity on plants is increase of the osmotic pressure in the root zone; this retards the absorption of groundwater by the plant roots.

In addition to the influence of osmotic pressure, the plant might suffer from the excessive concentration and excessive intake of ions which may be toxic for the plant. This may further inhibit the absorption of essential nutrients.

There is no definitive concentration limit of salinity with which plants are unable to grow. As salinity increases, however, plant growth is inhibited, developing chlorosis, and ultimately causing death. Tolerance to salt in soil is significantly different between plant species. Salt tolerance is evaluated by the difference between the yield in saline soil and that in non-saline soil.

Salinity	Category and notes
Less than 0.5 mS cm^{-1} Less than 250 μS cm^{-1} Less than 160 ppm Less than 160 mg L^{-1}	Low salinity water. In the case of improved saline soil, this type of water can be used as irrigation water for most types of agricultural land and for all kinds of crops.
0.7-2.25 mS cm^{-1} 250 - 750 μS cm^{-1} 160 - 480 ppm 160 - 480 mg L^{-1}	Brackish water. If appropriately leached, this type of water can be used for agriculture. Salt-tolerant plants can grow without special salinity adjustment.
0.75 -2.25 mS cm^{-1} 750 - 2250 μS cm^{-1} 480 - 1440 ppm 480 - 1440 mg L^{-1}	High salinity water. This type of water cannot be used for irrigation without drainage facilities. Land with enough drainage capacity can be used to grow only salt-tolerant plants under special salinity control.
More than 2.25 mS cm^{-1} More than 2250 μS cm^{-1} More than 1440 ppm More than 1440 mg L^{-1}	Extremely high salinity water. This type of water is inappropriate for irrigation water in general. However, it can be used in specific conditions. If soil has good permeability and high drainage capacity, only salt-tolerant plants can grow after leaching.

Table 1. Categories of salinity of irrigation water

If the soil in which the adverse effect is observed is mixed with water and then the water is analyzed, the salt concentration and the electrolytic conductivity (EC) will be more than 0.02 normality (N) and more than 2 mS cm^{-1}, respectively. When salt affects the growth of most plants, these parameters are likely to be more than 0.04 N and more than 4 mS cm^{-1}. In sodic and clay soil there is a deterioration in physical properties such as drainage capacity and air permeability, which result in growth inhibition.

Crops affected by salt exhibit lower physiological activity including photosynthesis and transpiration (see Table 1). According to the classification established by US Salinity Laboratory, soil with more than 7.5 mS cm^{-1} of EC significantly affect crops; only salt-tolerant plants can grow in this type of soil.

2.3 Salinity and osmotic pressure

Osmosis is the process of water movement from less saline water to more saline water through a semi-permeable membrane (Figure 1). The less saline water permeates the membrane to the more saline water until the salinities in both sides become balanced against the osmotic pressure which pulls water through the membrane. Semi-permeable membranes do not allow ions through them except water molecules. Theoretically, water can permeate the membrane to both sides, but more water molecules in the less saline side move to the more saline side because the amount of water molecules in the more saline side is smaller than that in the less saline side. As a result, the water level in the less saline side becomes lower while that in the more saline side becomes higher. The difference of the water levels pushes the less saline water. Ultimately, the differential pressure balances the salinities on both sides.

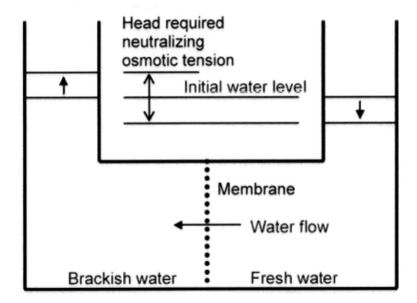

Figure 1. Osmosis and osmotic pressure

Because the positive pressure balances the salinities, the negative pressure is inversely generated on the less saline side and it pulls water outside the membrane. This negative pressure is called osmotic pressure. The osmotic pressure is also expressed as atmospheric pressure. The equation expressing the relationship between salinity and osmotic pressure is the following.

$$\text{Osmotic pressure (OP)} = \text{Salinity (mg L}^{-1}) \times 0.000563$$
$$= \text{Salinity (mS cm}^{-1}) \times 0.36 \qquad (1)$$

3. Salinity of water and soil

Salts in the irrigation water or soil solution cause an adverse effect of salinity on crop production. An index of salinity is the electrolytic conductivity (EC) that reflects the total salt concentration in the solution. EC is actually a measurement of electric current and is reported in units of milli-siemens per centimeter (mS cm^{-1}) or in micro-siemens per centimeter (μS cm^{-1}) at 25 degrees C (note: 1 mS = 1000 μS). Subscripts are used with the symbol EC to identify the source of the sample. ECiw is the electrical conductivity of the irrigation water. ECe is used for the saturated extract taken from the soil in the root zone and ECd for the saturated extract taken from below the root zone. ECd is used to determine the salinity of the drainage water that leaches below the root zone. In addition to EC, the concentration of salt may be measured or expressed in parts per million (ppm) or in the equivalent units of milligrams per liter (mg L^{-1}). If salts in a solution are not dissociated, such as $CaCO_3$, they do not contribute to EC. Only ions such as Na^+ or Cl^- that exist in free form contribute to EC. There is a negative relationship between EC and osmotic potential (OP) of a solution.

Problems that may arise from the use of salty irrigation water can be classified into two types. One type, which primarily affects the plant, is referred to as a salinity hazard that can lead to a saline soil condition. The other type affects the soil itself and is referred as a sodium hazard, which can lead to a sodic soil condition. In some cases there is a combined hazard, which results from both conditions and can lead to a saline-sodic soil condition.

3.1. Salinity hazard

Water availability of soils is expressed as a value of water potential (WP). The WP consists of matric potential (MP) and OP, excluding gravitational potential (GP). The wilting point of crops is −1.5 Mpa WP (minus 1.5 Mega Pascal in Water Potential), below which they cannot absorb water from the soil. In humid areas, soil WP is nearly equal to MP, i.e. the contribution of OP to WP is small. Excess salt increases the osmotic pressure of the soil solution and can result in a physiological drought condition. That is, even though the field appears to have plenty of moisture, the plants wilt because the roots are unable to absorb the water. In arid areas, as irrigation water usually contains a considerable amounts of salts, the contribution of OP is not negligible. As soil EC increases, soil WP reaches the wilting point at higher water content. From a viewpoint of water availability, to such an extent that yield is affected, there is a guideline for interpretation of irrigation water by FAO. If EC is lower than 0.7 mS cm^{-1}, there is no problem. If EC is between 0.7 and 3.0 mS cm^{-1}, there is a slight to moderate problem. If it is higher than 3.0 mS cm^{-1}, the problem is severe.

3.2. Sodium hazard

Salinity can also influence the infiltration rate of soil. Low saline water tends to leach soluble salts, especially calcium from the surface soils. Such soils lose stable aggregates and structure. They tend to disperse and the dispersed soil particles fill soil pore spaces, increasing the solid ratio and decreasing the porosity. In such soils, the irrigation water or rainfall remains on the soil surface and results in runoff. Another water quality factor that influences the infiltration rate is sodium concentration relative to calcium and magnesium. Sodium combines with clay particles and disperses them, which results in a decrease of infiltration of water into the soil. This factor is expressed by sodium adsorption ratio (SAR) in the following equation:

$$SAR = \frac{Na}{\sqrt{(Ca + Mg)/2}} \qquad (2)$$

where concentration of Na, Ca and Mg is expressed by cmol kg^{-1}.

The calcium and magnesium ions are important since they tend to counterbalance the effects of sodium. For water containing significant amounts of bicarbonate, the adjusted sodium adsorption ratio (SARadj) is sometimes used.

Continued use of water having a high SAR leads to a breakdown in the physical structure of the soil because the sodium absorbs onto the soil particles and the clay particles disperse. When the soil dries, it becomes hard and compact and increasingly impervious to water penetration. Finely textured soils, especially those high in clay, are most subject to this action. Gypsum can be economically used to improve some soils with high SARs. Calcium and magnesium, if present in the soil in large enough quantities, will counter the effects of the sodium and promote good soil properties.

Sometimes the soluble sodium percent (SSP) is used to evaluate sodium hazard. SSP is defined as the rate of sodium in epm (equivalents per million) to the total cation epm multiplied by 100. Water with an SSP greater than 60% may cause sodium accumulations that will break down the soil's physical properties.

At low leaching fraction (less than 0.1), precipitation of calcium with carbonate should be taken into consideration for near-surface soil.

The SAR is also used to evaluate sodium hazard (toxicity) to plants. Sodium is one of the most detrimental elements to crops. In most crops, sodium accumulates in the roots. A direct effect of sodium to the leaves, therefore, is unusual except for some woody species. In stone-fruit trees such as almond and peach, sodium is normally retained in the roots and lower trunk, but after three or four years, the conversion of sapwood to heartwood release the accumulated sodium, which is transported to the leaves causing leaf burn.

The indirect effect of sodium is more common. Sodium competitively inhibits the absorption of essential cations. An increase in the sodium concentration in the medium brings about a decrease in the potassium, calcium or magnesium concentration in crops, especially in the shoots.

Sodium in the irrigation water also contributes to increasing soil pH. Some essential micronutrients such as iron, manganese and zinc become less available under high soil pH and their deficiency symptoms can be seen in the leaves.

Some ions accumulate in the soil solution in amounts sufficient to cause toxic reactions in the plant. Sodium, chloride, bicarbonate and sulfate are most likely to cause problems. Crops grown in soils having an imbalance of calcium and magnesium may also exhibit toxic symptoms.

Sulfate salts affect sensitive crops by limiting the uptake of calcium and increasing the adsorption of sodium and potassium, resulting in a disturbance in the cationic balance within the plant. The bicarbonate ion in the soil solution harms the mineral nutrition of the plant through its effects on the uptake and metabolism of nutrients. High concentrations of potassium may cause a magnesium deficiency and iron chlorosis. An imbalance of magnesium and potassium may be toxic, but high calcium levels can reduce the effects of both.

4. Prevention of salinization and improvement of saline soil

The best way to prevent the salinization of soil is to use fresh water for irrigation. Water with approximately 0.25 mS cm^{-1} of EC can also be used for irrigation without causing salinization if the sodium concentration in the water is low. The allowable sodium concentration in irrigation water is determined by the amount of total water, the amount of underground drainage and the sodium adsorption ratio (SAR).

Irrigation water with less than 7.0 mmol$^{1/2}$ L$^{-1/2}$ of SRA is considered not to develop sodic problems. The alkalinization of soil is determined by residual sodium carbonate (RSC), which is the difference between the amount of carbonate (including bicarbonate) and the amount of calcium and magnesium in irrigation water. RSC of irrigation water less than 1.25 meq L^{-1} is considered unlikely to develop alkalinization.

Investigation of plant physiology in relation to salinization has made a great contribution. In this field, the transpiration and the structure and function of stomata, which absorb carbon dioxide, have been studied in association with photosynthesis. With this knowledge, the transpiration ratio is obtained, as an index of crop yield.

In terms of the absorption and movement of water by plants, the relationship of irrigation water and yield affects the relationship between the moisture condition of soil and transpiration ratio. This explains why salinity is an important factor for irrigation control.

During the process of osmosis, salinity generates a tension similarly to moisture tension. These tensions adversely affect plants. The yield of crops will be maximized if the total tension is kept at a lower level. The method of maintaining low moisture tension for crops differs between irrigation systems.

4.1. Prevention of capillary ascent of salts

In order to prevent the accumulation of salts in soil, the ascent of groundwater (and also dissolved salts) should be avoided. Mulching with plastic sheets or straw, mixing organic matter with surface soil, or plowing has been used to prevent the ascent of groundwater.

In recent years, the drip irrigation system has been widely applied in arid and desert regions and for afforestation. This system is useful not only to prevent salinization, but also to save important and limited water resources. In addition, this method also reduces damage by salinity in water and soil without causing the accumulation of salt. It supplies water equally and the automated operation can be applied easily.

The low-angle sprinkling method is also used. The ejection angle of sprinklers is set lower, thus avoiding the accumulation of salts in irrigation water on the surface of leaves, which can damage the plants. This type of sprinklers is mainly installed in orchards such as citrus trees.

The capillary effect of groundwater can be blocked completely by an asphalt water-blocking sheet or a plastic water-blocking film. Simultaneous use of such materials with drip irrigation reduces and prevents salt accumulation.

4.2. Leaching

In order to remove accumulated salts after plant damage occurs, rotational cultivation between rice paddy and upland crops, or just rice cultivation in a paddy field may be effective. During the period of cultivation with flooded water, the salts accumulated in the soil are leached out by the large volume of irrigation water that penetrates into the ground. The water must be drained away once in a few years in rotational cultivation and every year for rice growing.

The leaching method is a popular method to prevent salinization. More water is applied to the field than is required for crop growth. This additional water infiltrates into the soil and percolates through the root zone. During percolation; it takes up a part of the salts in the soil and takes these down to deeper strata. The water washes the salts out of the root zone. However, the leaching effect is affected by many factors, such as the degree of salinization, the type of saline soil, rock in soil, moisture, the depth of drainages, leaching procedure, and leaching time.

For sandy soil with extremely high water permeability, salt balance should be carefully measured as even the water contains less salt.

If water with high sodium content is used for irrigation, the ESP will become higher. Improvement of sodic soils requires reduction of the amount of sodium present in the soil. This is done in two stages. Firstly chemicals rich in calcium are mixed with the soil in order to replace the sodium with calcium. Then the replaced sodium is leached from the root zone by irrigation water.

Principally, four types of leaching procedures are practiced, depending on the amount of water required.

$$Q=Q_1+M-m \qquad (3)$$

$$Q=Q_1+Q_2 \tag{4}$$

$$Q=Q_1+Q_2+Q_3 \tag{5}$$

$$Q=R \tag{6}$$

Where,

Q: the amount of leaching water
Q_1: the difference between natural water content and water capacity in a field calculated by $Q_1=M-m$ where M expresses water capacity in the field and m expresses natural water content.
Q_2: the difference between water capacity in a field and that after saturation calculated by $Q_2=M-P$ where M expresses water capacity in the field and P expresses water amount after saturation.
Q_3: the amount of water penetrated into the ground after saturation calculated by the equation of $Q_3=nP$, or by multiplying water amount after saturation by the coefficient of n. The value of n varies depending on salt accumulation in the soil and physical properties of soil and water.
R: the amount of water lost from the ground (due to evaporation or penetration into the ground).

Equation (3) is used for irrigation of salt-accumulated soil. In this case, leaching and irrigation are conducted at the same time. The amount of salt in the soil is not changed. Leaching only carries salt into deeper soil.

Equation (4) is used to leach salt into groundwater. In this case, the salt balance in the whole area is not changed.

Equation (5) is used for land where water artificially or naturally flows into culverts.

Equation (6) is used for soil salt to ascend to the ground surface by water flow. After that, the separated soil salt is flushed away.

4.3. Required amount of leaching water

Estimation of the amount of water required to leach soil salt is especially important in improving soil salinity. The required amount is determined by the initial salinity in the soil, the desirable level of salinity after leaching, the depth of soil to be improved, and the type of soil. The amount of water required for leaching is called "leaching requirement (LR)", which is the proportion of the water required for leaching salt to the total amount of irrigation water, expressed by water depth unit. LR is determined from the EC of the irrigation water and the soil.

The following values can be used for rough estimation of leaching efficiency. Unit depth of water (indicate amount of water) is assumed to remove 80% of salts in soil at unit depth (indicate level in soil). In other words, 30 cm of water will remove 80% of the salts at 30 cm depth of soil. Or, to reduce salts at 60 cm depth of soil to 20% of the initial amount, 60 cm depth of water is necessary.

To estimate appropriate LR, the salt leaching test is desirable to obtain a leaching curve in a certain area. The leaching curve expresses the relationship between the ratio of actual salinity in soil to initial salinity (Sa/Sb) and the depth of leaching water per unit depth of soil (Dw/Ds).

The efficiency of leaching is decreased if water that percolated into the ground rises again to the surface by capillary action. The leaching water therefore needs to be simultaneously drained away.

4.4. Drainage

Irrigation is the best way to stabilize agricultural productivity where rainfall is not enough for the crops or it does not fall evenly. Excess irrigation can, however, cause the ascent of groundwater. Once the groundwater level comes close to the ground surface, it starts to ascend due to evaporation from the ground, resulting in accumulation of salt around plant roots (see Figure 2). Generally, the limit of ground water level is approximately 1.5 to 3.0 m depending on the soil type, the depth of the plant roots and the salinity of the groundwater.

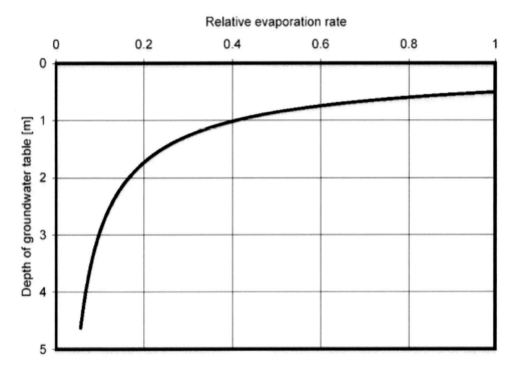

Figure 2. Relationship between depth of groundwater and relative evaporation rate of water on ground surface

Opened drainage ditches are more efficient than closed drainage ditches for land where salt has been accumulated in soil. Drainage systems vary depending on soil types. For example, in a drainage system for a slightly inclined field (slope less than 0.2%), a drain gutter is constructed along one side of the field, along with V-shaped shallow drainage ditches to drain water into the gutter.

A closed underground drainage system will be needed if excessive water and dissolved salt cannot be naturally drained and the groundwater level rises, as this adversely affects ventilation in the root zone.

4.5. Prevention by salt-tolerant plants

Mitigation measures for salinization, such as reducing soil salinity and cultivating salt-tolerant crops have long been practiced in saline land. Today, however, new varieties of crops with high salt-tolerance and higher salt assimilating capacity are being developed with the help of biotechnology.

Recently, attempts have been made to reuse wastewater or to use brackish water in order to utilize water resource more efficiently. Brackish water contains 1500 to 7000 mg L^{-1} of salts, naturally or artificially dissolved into the water (see Table 2).

Water Category	Electric Conductivity (mS cm^{-1})	Salinity (mg L^{-1})	Water Type
Fresh water	< 0.7	< 500	Drinking water, irrigation water
Low salinity	0.7 - 2	500 - 1500	Irrigation water
Brackish	2 - 10	1500 - 7000	Primary wastewater, groundwater
High salinity	10 - 25	7000 – 15 000	Secondary wastewater, groundwater
Extreme high salinity	25 - 45	15 000 – 35 000	Groundwater with very high salinity
Salt water	> 45	> 35 000	Seawater

Table 2. Classification of salinity in water

The salinity of surface water, wastewater, or groundwater is calculated from the total concentration of dissolved inorganic ions such as Na, Ca, Mg, K, HCO$_3$, SO$_4$ and Cl. In addition to the total salinity, Na and pH adversely affect the soil. Excessive Na affects divalent cations and clay minerals in soil. The effects become especially significant when total salinity and pH are higher. Proper analysis should be conducted to ascertain whether the salinity is caused by sodium ion or other salts, because high salinity may improve soil permeability.

The following remarks are guidance for use of brackish water for agricultural production.

- The most suitable salt-tolerant crops should be chosen.
- Control of seedbeds and improvement of fields should be performed in order to avoid partial salt accumulation.
- Soil properties should be deliberately controlled for irrigation.
- Drainage systems should be improved and used efficiently.
- An efficient and effective irrigation system should be chosen.

When brackish water is used for irrigation, an appropriate amount of water should be used in an appropriate manner. Selection of the irrigation method is a key to prevent salinization.

Glossary

CEC:	Cation Exchange Capacity.
EC:	Electrolytic Conductivity.
ESP:	Exchangeable Sodium Percentage.
FAO:	Food And Agriculture Organization.
GP:	Gravity Potential.
LR:	Leaching Requirement.
MP:	Matric Potential.
OP:	Osmotic Potential.
RSC:	Residual Sodium Carbonate.
SAR:	Sodium Adsorption Ratio, which is derived from the ratio of sodium and the square root of the average of calcium and magnesium.

Bibliography

Bohn H. L., McNeal B. L. and O'Connor G. A. (1985). *Salt-Affected Soils*, Soil Chemistry 2nd Ed., 234 – 261. New York: WILEY-INTERSCIENCE. [The Terminology Committee of Soil Science of America recommended lowering the boundary because salt-sensitive plants can be affected in soil whose saturation extracts have EC's of only 2 to 4 dS m^{-1}.]

Brouwer C., Goffeau A. and Heibloem M. (1985). *Irrigation water management training manual* No.-1 introduction to irrigation, Rome: Food and Agriculture Organization. [This paper is intended for use by field assistants in agricultural extension services and irrigation technicians at the village and district levels who want to increase their ability to deal with farm-level irrigation issues.]

FAO (1984) Water Quality Evaluation. Water Quality for Agriculture. *FAO Irrigation and Drainage Paper* 29 Rev. 1, 1 – 11. New York: Food and Agriculture Organization. [A guideline for water users such as water agencies, project planners, agriculturalists, scientists and trained field people to better understand the effect of water quality on soil conditions and crop production.]

Hagin J. and Tucker B. (1982). *Major Soil Characteristics.* Fertilization of Dryland and Irrigated Soils. 4 – 9. Berlin: Springer-Verlag. [Soils of arid and semi-arid regions which are classified from a viewpoint of salinity and sodicity in relation to their pH.]

Maas E. V. (1987). *Salt Tolerance of Plants.* Handbook of Plant Science in Agriculture Volume . 57 – 75. Florida: CRC Press. [Tolerance of crop species to salinity, chloride, sodium and boron is discussed by yield and injury.]

Suarez D. L. (1981). Relationship between pHc and SAR and an alternative method of estimating SAR of soil or drainage water. *Soil Sci. Soc. Am. J.*, 45, 469 – 475 [A new SAR with a modified calcium value in the water modified due to salinity of the water, its HCO_3/Ca ratio and the estimated partial pressure of CO_2 in the surface few millimeters of soil.]

U.S. Salinity Laboratory Staff (1954a). Determination of the Properties of Saline and Alkali Soils, Diagnosis and Improvement of Saline and Alkali Soils. 7 – 33. *Agriculture Handbook No. 60*, Washington: USDA. [A ratio for soil extracts and irrigation water used to express the relative activity of sodium ions in exchange reactions with soil.]

U.S. Salinity Laboratory Staff (1954b). Sampling, soil extracts, and salinity appraisal. Diagnosis and Improvement of Saline and Alkali Soils. *Agriculture Handbook No. 60*, USDA, Washington.

Biographical Sketches

Hideyasu Fujiyama is Professor of Agriculture at Tottori University, where he has been on faculty since 1996. He was admitted to Osaka University in 1969 and received the Bachelor Degree of Pharmacy in 1973. Then, he continued his research at Hokkaido University and obtained the Master Degree in Agriculture in 1976. From 1976, he served as a Research Associate at the Faculty of Agriculture, Tottori University, then as Associate Professor from 1989. A Doctoral Degree from Hokkaido University was conferred on him in 1987.

His domain of research is salinity and eutrophication; it includes the difference in mechanism of salinity hazards between plant species, the growth improvement of salt-stressed plants by chemical methods, the effect of acid rain on nutrient absorption and growth of plants, the dynamics of elements in water and sediments in eutrophic lakes, and the remediation of eutrophic lakes using the nutrient absorption ability of plants.

Yasumoto Magara is Professor of Engineering at Hokkaido University, where he has been on faculty since 1997. He was admitted to Hokkaido University in 1960 and received a degree of Bachelor of Engineering in Sanitary Engineering in 1964 and Master of Engineering in 1966. After working for the same university for four years, he moved to the National Institute of Public Health in 1970. He served as the Director of the Institute. From 1984 he worked for the Department of Sanitary Engineering, then the Department of Water Supply Engineering. He obtained a Ph.D. in Engineering from Hokkaido University in 1979 and was conferred an Honorary Doctoral Degree in Engineering from Chiangmai University in 1994. Since 1964, his research subjects have been in environmental engineering and have included advanced water purification for drinking water, control of hazardous chemicals in drinking water, planning and treatment of domestic waste including human excreta, management of ambient water quality, and mechanisms of biological wastewater treatment system performance. He has also been a member of governmental deliberation councils of several ministries and agencies including Ministry of Health and Welfare, Ministry of Education, Environmental Agency, and National Land Agency. He performs international activities with JICA (Japan International Cooperation Agency) and World Health Organization. As for academic fields, he plays a pivotal role in many associations and societies, and has been Chairman of Japan Society on Water Environment.

Professor Magara has written and edited books on analysis and assessment of drinking water. He has been the author or co-author of more than 100 research articles.

WATER POLLUTION BY AGRICULTURE AND OTHER RURAL USES

Yuichi Fushiwaki
Kanagawa Prefectural Institute of Public Health, Chigasaki, Japan

Yasumoto Magara
Hokkaido University, Sapporo, Japan

Keywords: agriculture, degradation, degradation by-products, dioxin, drinking water, nitrate, pesticide.

Contents

1. Introduction
2. Pesticides
3. Dioxin group
4. Nitrate
Glossary
Bibliography
Biographical Sketches

Summary

Contemporary agriculture heavily depended on chemical substances in order to achieve the "Green Revolution". Natural organic matter or compost was rapidly replaced by chemical fertilizers, and pesticides were widely used to secure the highest yield of desired crops. Pesticides are sprayed in paddy fields or crop fields for agricultural activities worldwide. In general, only a small part of the pesticide is effectively applied to the farm products; most of it enters the soil.

As a consequence of their persistency and bioaccumulation properties, certain pesticides, which are able to accumulate in higher organisms (even in human beings), have caused serious problems to ecosystems. Production and use of these pesticides have therefore been restricted in many countries. The properties of most pesticides have been changed so that they degrade in the environment.

During the production process of agricultural chemicals, trace amounts of dioxins are also produced, and these can remain in the final products. The dioxin group is not volatile and is poorly soluble; furthermore, they are very persistent in environment. As a result, they are adsorbed to particulate matter and finally settle in soil or sediment of rivers, lakes, or seas.

Nitrate is a relatively non-toxic substance which occurs naturally as part of the nitrogen cycle. Once ingested, however, nitrate is reduced to nitrite by bacterial activities. In the blood, nitrite oxidizes iron in the hemoglobin of red blood cells to form methaemoglobin, which lacks haemoglobin's oxygen-carrying ability.

Many water sources, at which high concentration of nitrate and nitrite have been reported, are often surrounded by orchards or vegetable farms, which implies water contamination by agricultural fertilizers. If nitrate (and nitrite) nitrogen levels of source water exceed 10 mg L^{-1}, some treatment process must be provided before the distribution of drinking water. However, no simple method to remove nitrate has been developed. Usually, contaminated water is mixed with other non-polluted water sources before being supplied for drinking purpose.

1. Introduction

Contemporary agriculture heavily depended on chemical substances in order to achieve the "Green Revolution". Natural organic matter or compost was rapidly replaced by chemical fertilizers, and pesticides were widely used to secure the highest yield of desired crops. Pesticides are sprayed in paddy fields or crop fields for agricultural activities worldwide. Most of the pesticide applied is degraded in agricultural fields, if it is applied properly, and does not cause serious water pollution. In general, however, only a small part of the pesticide is effectively applied to the farm products; most of it enters the soil. In addition, some pesticide scatters during spraying and is carried outside the fields.

In recent years, more attention has been paid to environmental pollution caused by chemical substances, and their adverse effects on the environment, including those on human health, have been much discussed. Pesticides and other agricultural chemicals are different from other chemicals because they are used in open areas and in large quantity, and are used only in certain seasons. Therefore, environmental pollution by agricultural chemicals needs to be carefully considered.

The behavior of pesticides in the environment can be summarized by their volatility, solubility in water, partition coefficient in water and air, adsorption, biodegradation and concentration. These characteristics are so diverse that the pesticides sprayed on the fields move between the compartments of air, water and soil, and thus the fates of the pesticides are determined by their particular properties.

Chemical fertilizers also affect the environment. Excessive application causes eutrophication of environmental water, and can cause nitrate contamination, which is a health hazard in drinking water. Several measures are proposed to control and reduce nitrogen level in environmental water.

2. Pesticides

2.1. Behavior of pesticides in the atmosphere

A part of the pesticides which falls on the soil, or which has been directly applied to the soil, evaporates and passes into the atmosphere. As a result, a certain amount is transferred into the atmosphere as gas or in a form attached to suspended particulates. In real terms, however, the pesticides in the atmosphere are diffused and diluted by various atmospheric phenomena such as movement of air masses and fluctuation of wind direction. Pesticide pollution in the air is, therefore, usually a local and temporary phenomenon. Although the pollution level is high immediately after spray of pesticides, the level usually drops in a few days. Air pollution may continue for a longer period if a large amount of pesticides is sprayed over wide areas.

Pesticides in the atmosphere are slammed down with the falling rain and pass into the soil and thus, the water system. Not only falling rain but also photochemical reactions contribute significantly to the reduction of the pesticides in the atmosphere. Photolysis of pesticides is essentially the same as the photochemical reaction in organic chemistry. For example, an organic phosphorus insecticide, parathion, is known to be photo-oxidated from a thiono compound (P=S) to an oxon derivative (P=S double bond is replaced by P=O double bond), as shown in Figure 1, by ultraviolet irradiation. In addition, isomerization of the chemical to thiol derivative (P-S-) is also recognized. Some of these photolytic products have stronger residual capability and toxicity than the original chemical compounds. Therefore, attention should also be paid to the behavior of these photolytic products.

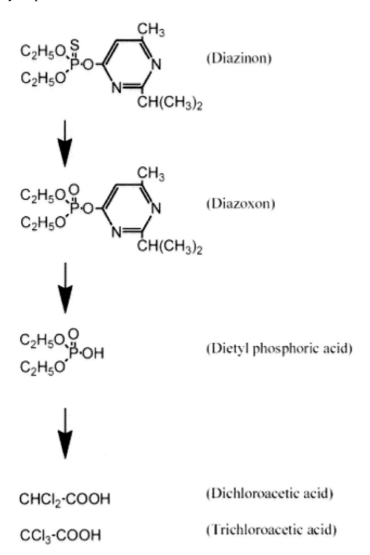

Figure 1. Degradation of diazinon by photolysis or chlorination

2.2. Run-off rate of pesticides from rice fields

Pesticides in the hydrosphere are present in various forms depending on their solubility in water and their physicochemical properties. In general, the level of pesticides in streams gradually lowers due to biodegradation and dilution. The runoff of pesticides into rivers varies with the amount and the duration of pesticide use.

In order to identify the quantity of pesticides flowing into rivers, a study was implemented in a catchment area of about 86 km^2, 10% of which was occupied by paddy fields. A water intake was located at 13 km downstream from the head of the river, and average river flow was 1.6 to 2.0 m^3 sec^{-1}. Agricultural irrigation water was collected to determine the run-off rate of pesticides used in the basin area. Water samplings were carried out twice a month from May to November.

	Year	1	2	3	4
	Pesticide name	Concentration (µg L^{-1})			
Insecticide	diazinon	0.03-0.21	N/D	0.013-0.066	0.133-0.190
	fenitrothion	N/D	0.002-0.26	0.046-0.284	0.546
	isoxathion	N/D	0.15-3.33	N/D	0.749-6.76
	chlorpyrifos	N/D	0.05-0.14	N/D	N/D
Germicide	iprobenfos	5.15-8.64	N/D	0.009-1.089	0.002-1.43
	isoprothiolane	0.72-69.83	0.01-2.87	0.057-9.617	0.028-30.00
	fthalide	0.05-4.16	0.02-1.33	N/D	0.004-22.60
	flutolanil	0.37-20.78	0.55-36.52	0.005-3.412	0.039-137.0
	captan	N/D	0.07-0.27	N/D	N/D
	tolclofos-methyl	N/D	N/D	0.003	N/D
Herbicide	oxadiazon	0.11-1.05	0.01-0.84	N/D	0.018-0.242
	butachlor	0.29-41.70	0.2	N/D	N/D
	chornitrofen	1.213	N/D	0.012-0.141	N/D
	chlormethoxynil	1.569	0.04-0.14	N/D	N/D
	bromobutide	0.41-4.73	N/D	0.003-1.171	0.008-0.648
	thiobencarb	0.23-15.09	0.02-0.07	0.005-14.17	0.015-1.69
	esprocarb	0.23-9.47	N/D	0.006-1.096	0.004-0.192
	simazine	N/D	N/D	0.007-0.078	N/D
	prometryn	0.07-3.02	N/D	N/D	N/D
	simetryn	0.11-0.87	N/D	0.005-0.655	N/D
	dimethametryn	N/D	0.01-0.02	0.017	N/D
	pretilachlor	N/D	0.05-0.07	0.065-0.132	N/D
	bifenox	N/D	0.02-0.48	N/D	N/D
	propyzamide	N/D	0.02	N/D	0.026
*	isoxathion-oxon	N/D	N/D	N/D	0.059-23.90
	chlorpyrifos-oxon	N/D	N/D	N/D	0.027-72.00

Note: * Degradation by-product of pesticide

Table 1. Residual pesticides in water

Table 1 shows the results of a survey of pesticides in environmental water (agricultural waterways) in four consecutive agricultural years. In these four years, twenty-four kinds of pesticides (four kinds of insecticides, six kinds of germicides, fourteen kinds of herbicides) were detected at a level higher than the detection limit. However, only isoprothiolane, flutolanil and thiobencarb were detected in all four consecutive years. These results show that the types of pesticides applied to fields differ depending on the conditions of climate or agricultural management.

In addition, two kinds of pesticide degradation product, i.e. isoxathion and chlorpyrifos-oxon, were detected. The relationship between solubility and maximum concentration was shown in Figure 2. It appeared that the plots in Figure 2 were classified into two groups. But the correlation was observed that the maximum concentrations increased with increase of solubility.

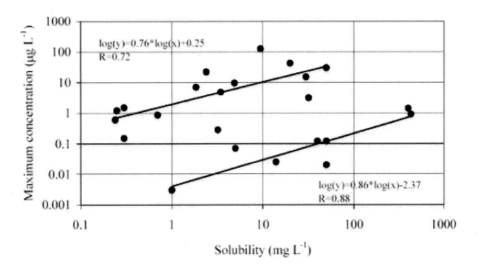

Figure 2. Relation between solubility and maximum concentration

The run-off rate of pesticides was estimated from the flux of pesticides in river water and the shipment amounts of pesticides in surveyed area. The flux of pesticides was estimated from average concentration of pesticides per sampling point and rate of river flow every day. The run-off rates of the main pesticides were calculated as shown in Table 2. Thiobencarb was about 3%, fthalide, iprobenfos and oxadiazon ranged from 18 to 35%, and flutolanil was 80%. The run-off rate of fthalide, iprobenfos and oxadiazon showed fairly high values in comparison with thiobencarb or esprocarb.

Name	Shipment amount (kg)	Run-off amount (kg)	Run-off rate (%)
iprobenfos	12.1	3.61	29.8
thiobencarb	273	8.807	3.2
esprocarb	693.9	19.38	2.8
fthalide	602.6	109.7	18.2
flutolanil	368.3	297.9	80.9
oxadiazon	18.2	6.433	35.3

Table 2. Estimation of run-off of pesticides from agricultural area

It has become evident that soil particles containing pesticides flowed into rivers from the soil surface with rainfall and increased the pesticide load. Pesticides not only effuse from the surface of the soil but also percolate downward through the soil and gradually flow into the river. Part of the pesticides percolate down to the groundwater and leaves the rice paddy areas in downward drainage. Therefore, experiments have been carried out using a lysimeter to elucidate the movement and effluence of herbicides by the percolation of water through the soil. According to the survey, the amount of the effluence of CNP (herbicide) was smaller than that of benthiocarb, simetryne and molinate. The amount of herbicides that percolated downward through the soil was reportedly the highest in molinate, followed by simetryne, benthiocarb and CNP. Thus, the amount of percolation was larger in pesticides with higher solubility in water.

It is considered that fish and shellfish are contaminated with pesticides that concentrate in their bodies. These come from pesticides sprayed in paddy and crop fields that flow into agricultural waterways and discharge into rivers and finally into the sea. It is also a matter of concern that pesticides accumulate in higher animals via the food chain.

2.3. Behavior in soil

The behavior of pesticide in the soil is mostly determined by biodegradation, metabolism, adsorption of organic matter and clay mineral in the soil, and chemical decomposition. These transformations of pesticides in the soil vary with the type and abundance of microorganisms, the nature of the soil, the composition of clay minerals, the content of organic matter in the soil, and pH. The rate of biodegradation and the process of metabolism greatly differ between aerobic and anaerobic conditions. For example, benthiocarb is biodegraded rapidly under aerobic conditions but slowly under anaerobic conditions. Pentachlorophenol (PCP) is completely biodegraded through the cleavage of its benzene ring by microorganisms in the farm soil. The fungicide, PCNB, undergoes aerobic degradation slowly, but anaerobic degradation is very rapid. The biological conversion of the nitro group into the amino group by the microbial reducing process proceeds rapidly. On the other hand, no degradation was reported with benthiocarb after 40 days under anaerobic conditions. In addition, certain pesticides and their degradation products firmly adhere to clay minerals and humic materials in the soil, and they show a high residual rate. However, few studies so far have been performed on environmental toxicology of these degradation products. This is an issue that deserves more thorough elucidation. It has become clear that reductive dechlorination of the benzene ring of benthiocarb, etc. is affected by the activities of anaerobic microorganisms. It is supposed that these chemicals are subjected to cleavage of the benzene ring; they are then metabolized to fatty acids, and finally become inorganic carbon. The half-life of pesticide in the soil depends on the soil condition and the spray condition, but in general, it ranges from three days for the shortest up to about three months for the longest.

Pesticides sprayed in paddy and crop fields are adsorbed onto soil particles and are gradually degraded biologically. However, it has become evident that some parts of the pesticides remain in the soil. As pesticides also leach into rivers, the environmental impacts of these substances in rivers must be carefully considered. Some pesticides and degradation products show great bioaccumulation capability, and they are detected also in source water during the season when they are applied. The adverse effects of the pesticides on human health through the food chain are also a matter of concern due to their high bioaccumulation. Therefore, it is necessary to pay attention to pesticide pollution and make further efforts to take appropriate countermeasures.

2.4. Degradation by-products in water purification system

Pesticides which are applied to agricultural fields leach to the water environment, as described above. In order to anticipate proper control of pesticides in drinking water it is necessary to know their behavior not only in the environment but also in the water purification process. It is considered that chlorination and ozonation, which are the principal processes of water purification, can produce by-products from reaction between chlorine or ozone and pesticides in source water. It has been reported that in the case of some organophosphate pesticides, the degradation byproducts have higher toxicity than the original pesticides.

An experiment involving chlorination of pesticides was performed, and the chlorination by-products were identified by GC-MS as diazinon from organophosphate insecticides, and thiobencarb from carbamate type herbicides. These pesticides are used in large quantity, and appear at high concentration and frequency in rivers. For quantitative determination of thiobencarb and its chlorination by-products, HPLC with a photodiode array detector was applied, and degradation by-products were surveyed on the filtrate of rapid sand filter of an actual purification plant.

Diazinon, which is typical of organophosphate pesticides, which have a P=S bond, was chlorinated to identify the degradation path and the by-products in a laboratory. Chlorination of diazinon resulted in the detection of diazoxon, which is an oxon. When the reaction between diazoxon and chlorine is continued, the degradation develops. Oxidation by chlorination or hydrolysis turned to the by-products into hydrophilic compounds with high polarity, such as carboxylic acid. Ethyl-acetate-extracted solution showed three types of by-products, namely dichloromethyl acetate, trichloromethyl acetate and methyldiethyl phosphate. From these results, it is confirmed that chlorination degrades diazinon into diethyl phosphate through oxons, and into dichloro acetate and trichloro acetate as shown in Figure 1. The stability of degradation by-products of diazinon in water is so high that 40 to 50% of diazoxon; for example, remained even after 48 hours of reaction time, and with chlorine concentration at 5 mg L^{-1}.

The paths of degradation by chlorine of these organophosphate pesticides are shown in Figure3. Dithio and thiono substances can easily be turned by chlorination into oxon through the P=O bond, then into phosphoric acid and other substances (R-OH). Thiol and phosphonate substances that have a P=O bond in their molecules are also degraded into phosphoric acid and other substance (R-OH). However, phosphonate substances, such as DDVP, which do not have a sulfur atom, are not easily decomposed by chlorine.

Thiobencarb, which is a typical insecticide in rice farming, is also chlorinated to become chlorination by-products in water purification system. When performing chlorination of thiobencarb, 80% or more was degraded immediately. In about two hours, the entire thiobencarb was degraded. The degradation products were identified by GC-MS, as chlorobenzyl alcohol, chlorotoluene, chlorobenzyl chloride, chlorobenzoic acid and chlorobenzyl aldehyde. The degradation process of thiobencarb through chlorination is believed to be as shown in Figure 4.

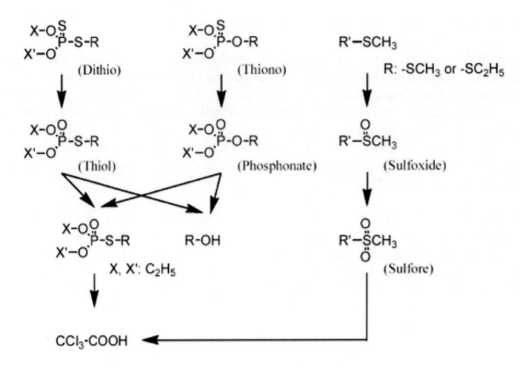

Figure 3. Degradation of organophosphate by chlorination

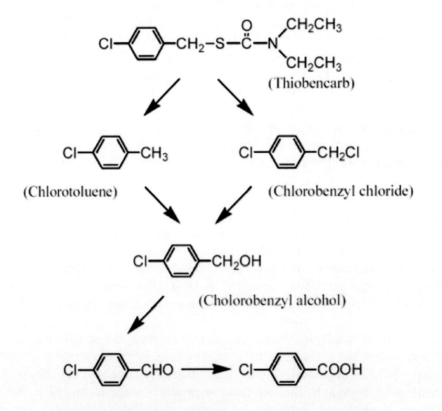

Figure 4. Degradation of thinobencarb by chlorination

The concentration of chlorination by-products of thiobencarb was studied in tap water during the season of its application in rice fields. In the filtrated water subjected to prechlorination, the degradation by-products of thiobencarb were detected. In particular, the concentration of chlorobenzyl chloride was the highest (about 12µg L^{-1}).

It is concluded that many kinds of pesticides may be degraded by chlorination since chlorine is a strong oxidant that does not exist under natural conditions. Therefore, the management and control of pesticides in drinking water and environmental water quality management should include consideration of chlorination by-products.

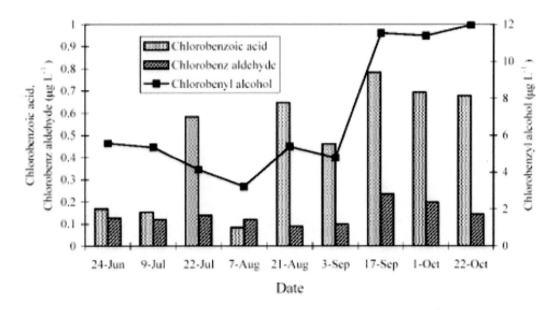

Figure 5. Degradation by-products in tap water

3. Dioxin group

3.1. Characteristics of dioxin group

Poly-chlorinated dibenzo-dioxin (PCDD) and poly-chlorinated dibenzo-furan (PCDF) are both chlorinated hydrocarbons which have been found to be among the most toxic chemical substances that human beings have ever created. Both PCDD and PCDF have many congeners (up to 210) depending on the location and number of chlorine atoms attached to the main structure. Collectively, they are called "dioxins" and "furans" or in short, "dioxin group". General structures of PCDD and PCDF are shown in Figure 6.

Well-known sources of dioxin group are emission from incinerators, and effluent from pulp and paper mill industries. In addition, during the production process of agricultural chemicals, trace amounts of dioxin group are also produced, and these remain in the final products. The dioxin group is not volatile and is poorly soluble; furthermore, they are very persistent in environment. As a result, they are adsorbed to particulate matter and finally settle in soil or sediment of rivers, lakes, or seas.

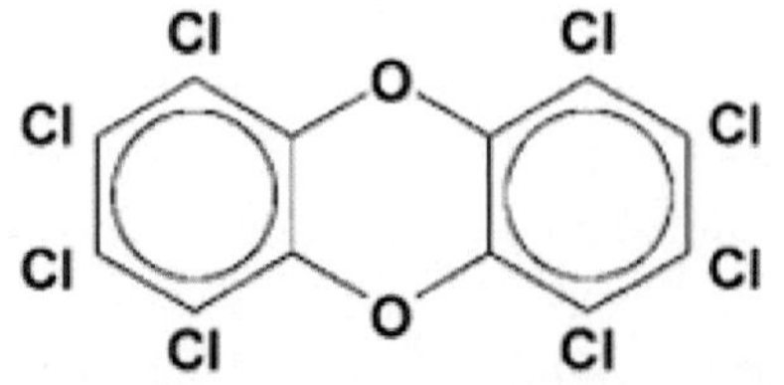

Polychlorinateddibenzodioxin (OCDD)

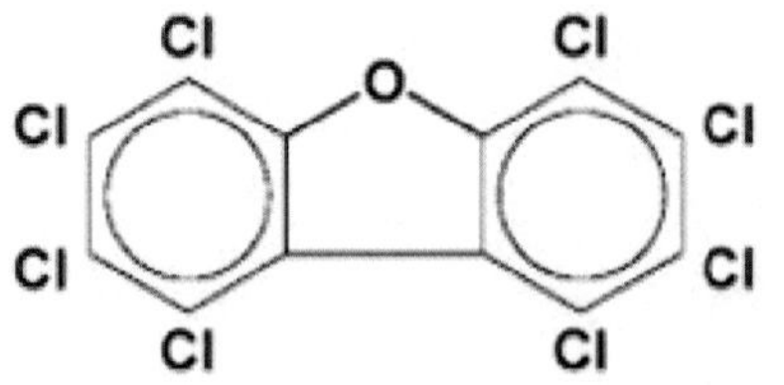

Polychlorinateddibenzofuran (OCDF)

TeCDD, TeCDF : Dioxin or furan congeners with four chlorine atoms out of eight replacement sites
PeCDD, PeCDF : Dioxin or furan congeners with five chlorine atoms out of eight replacement sites
HxCDD, HxCDF : Dioxin or furan congeners with six chlorine atoms out of eight replacement sites
HpCDD, HpCDF : Dioxin or furan congeners with seven chlorine atoms out of eight replacement sites
OCDD, OCDF : Dioxin or furan congeners with eight chlorine atoms occupying all replacement sites

Figure 6: Chemical structure of dioxin group

3.2. Impurity in agricultural chemicals

Pentachlorophenol (PCP) and chloronitrophen (CNP) are both chlorinated hydrocarbons used as pesticides in agriculture and forestry. During the production of PCP and CNP, unexpected reactions occur to produce unwanted by-products, particularly dioxin group. Due to their particular structures, PCP and CNP contain congeners of dioxins and furans. Figure 7 and 8 show the results of chemical analysis of the impurities.

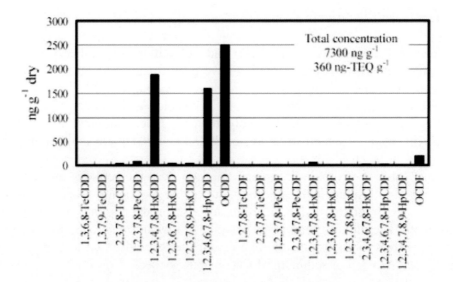

Figure 7. Dioxin impurities in PCP

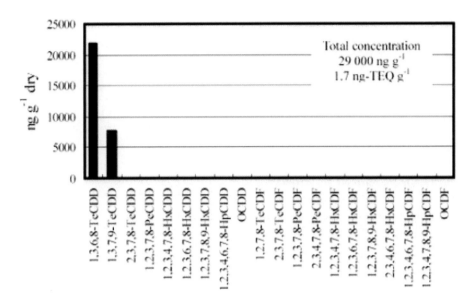

Figure 8. Dioxin impurities in CNP

3.3. Contribution of agricultural activity to dioxin concentration in environmental water

Since PCDD and PCDF comprise 210 congeners, the analysis of these compounds yields a considerable amount of data. Multivariate analysis has been utilized to reduce the complexity of the data.

Ten sampling sites in Japan with dioxin concentrations higher than 70 pg L^{-1} were considered. Samples were collected three times in each site and 83 out of 210 congeners were identified. The range of data was analyzed by a multivariate method, namely factor analysis. Table 3 shows the result and interpretation of the extracted factors.

Factor	Contribution	Cumulative contribution	Characteristic congeners (Factor loading > 0.7)	Interpretation
Factor 1	0.34	0.34	Some TeCDFs and PeCDDs, most of PeCDFs and HxCDFs	Combustion
Factor 2	0.24	0.58	Some TeCDDs, PeCDDs, and 2468-TeCDF, especially 1368, 1379-substituted TeCDDs and TeCDFs	CNP
Factor 3	0.18	0.76	Some HxCDDs and HxCDFs, OCDD, OCDF	PCP
Factor 4	0.06	0.82	Some TeCDDs, and TeCDFs	Uncertain

Table 3. Result of factor analysis in environmental water

From the significant congeners in each factor, possible sources of the contaminants were estimated. Factor 1 was interpreted as combustion origin, which contributed 34% of total dioxin concentration. The contribution of CNP and PCP accounted for 24% and

18%, respectively, which ranked as second and the third places in the contributions. In total, the contribution of CNP and PCP became 42%, exceeding the contribution of combustion.

Ten of the sampling sites in Japan are rather agricultural areas. For these regions, the main contributors of dioxin contamination were agricultural activities.

4. Nitrate

4.1. Health effects caused by nitrate

Nitrate is a relatively non-toxic substance which occurs naturally as part of the nitrogen cycle. Ammonium is excreted by animals and is taken up by plants to produce amino acids and other nitrogen-containing compounds. There is another type of microorganism which uses ammonia as an energy source; oxidizing it into nitrate under aerobic conditions. As a result, nitrate remains in non-contaminated water. Once ingested, nitrate is reduced to nitrite by bacterial activities. In the blood, nitrite oxidizes iron in the hemoglobin of red blood cells to form methemoglobin, which lacks hemoglobin's oxygen-carrying ability.

Although methemoglobin is continually produced in humans, an enzyme in the human body reduces it to hemoglobin. In most individuals, methemoglobin is rapidly converted back to hemoglobin, but infants have a low concentration of the reducing enzyme, as do some older individuals with an enzyme deficiency. In these people, methemoglobin is not as readily converted to hemoglobin.

When methemoglobin levels are elevated, a condition known as methemoglobinemia, often referred to as "blue baby syndrome", can result, as the blood lacks the ability to carry sufficient oxygen to individual body cells.

4.2. Nitrate pollution from agricultural activities

Many water sources in which high concentrations of nitrate and nitrite have been reported, are often surrounded by orchards or vegetable farms, which implies water contamination by agricultural fertilizers. Excreta from cattle and domestic wastewater are also regarded as possible contamination source of nitrogen. Contaminated water is mixed with another non-polluted water source before being supplied for drinking purpose.

Table 4 shows the result of multi-regression analysis of nitrate contamination in source water. Population density, annual precipitation, chemical fertilizer consumption per unit surface and excrement of cattle per unit surface were assigned as the parameters of the analysis. Though multi-regression coefficients of all data were not very high, they were all significant at 1% by T-test. The regression coefficient of population density or livestock excreta was not significant, but chemical fertilizer was significant in all analyses. It is clear that chemical fertilizers play an important role in the nitrate concentration of source water.

| Type | Constant | Regression coefficient | | | | Data number | Multi-regression coefficient |
		Population density [p km-2]	Annual precipitation [mm y^{-1}]	Chemical fertilizer [Nt km^{-2} y^{-1}]	Cattle manure [Nt km^{-2} y^{-1}]		
Total	1.34	4.78E-5 (-)	-5.99E-4 (**)	3.39E-1 (**)	4.35E-2 (-)	250	0.53 (**)
Shallow well	2.42	1.60E-4 (-)	-9.13E-4 (**)	2.37E-1 (**)	1.74E-2 (-)	98	0.49 (**)
Deep well	0.05	-1.02E-5 (-)	-1.90E-4 (-)	4.43E-1 (**)	3.43E-2 (-)	114	0.63 (**)

Note: Annual average value of nitrate (+ nitrite) concentration was applied
        ** is significant at 1% by t-test

Table 4. Multi-regression analysis of nitrate contamination in source water

4.3. Measures against nitrate contamination

4.3.1 Control of fertilizers

Fertilizers need to be developed to improve the uptake rate of nitrogen by crops, and to reduce residual nitrogen in soil, that has not been utilized by the crops. The amount, timing and methods of application of the fertilizers are all relevant in minimizing the leaching of nitrogen.

In paddy field, the application of fertilizer was halved at the initial stage of the plantation, which reduced the nitrogen concentration by 75%. However, it caused poor growth of rice, so a small amount of fertilizer was then added and the crop yield was increased by a few percent."

Excessive amounts of fertilizer tend to be applied to vegetable fields. Reduction of fertilizer does not affect the yield of some crops, and it effectively improves the nitrogen uptake rate. Especially, salt-tolerant crops such as strawberry or green onion are considered to be unable to absorb fertilizer efficiently at low salt concentration in soil. Therefore, excessive application of fertilizer is often practiced and this increases leaching of nitrogen. For those crops which have longer cultivation periods, the amount of residual nitrogen in soil can be reduced by repeatedly applying a small amount of fertilizer, or by applying fertilizer along with the crops instead of complete mixing in soil.

In Japan, water-soluble substances such as ammonium sulfate, ammonium chloride and urea are often used as nitrogen source. Ammonium bases dissociated in soil solution (urea is also hydrolyzed into ammonia by urease in soil) first adhere on the surfaces of soil particles by negative charge. Then, they are quickly oxidized to nitrate by nitrifying bacteria in soil, and leached into water. It is therefore important that the fertilizer should remain adhered onto the soil surface, and to control the activity of nitrifying bacteria in the soil. Fertilizer with larger particle size or coated by a synthetic resin physically retards the dissolution of the nitrogen component. Another method is to use substances with low solubility as nitrogen component, e.g. urea derivatives, dicyandiamide derivatives or oxamide. Nitration inhibitors are fertilizers additives which control the activity of nitrifying bacteria. Originally, these fertilizers were developed in order to improve the uptake rate of nitrogen in view of the productivity of crops. They are also effective, however, in slowing the release of nitrate nitrogen.

It is known that organic fertilizer also prevents the release of nitrate nitrogen. One reason is that the organic nitrogen takes longer time to be oxidized to nitrate because the reaction is a multi-stage process. Another reason is that higher a denitrification rate is expected in fields on which organic fertilizer is applied because anoxic conditions tend to occur because of the decomposition of organic matter. A study was conducted to evaluate the denitrification capacity of organic matter. Results were compared between an area in which 370 kg ha^{-1} of chemical fertilizer was applied and another area in which compost (equivalent to 292 kg ha^{-1} of nitrogen) was applied in addition to 370 kg ha^{-1} of chemical fertilizer. The amount of nitrogen leached out of the first experiment was 85 kg ha^{-1} while the latter was only 66 kg ha^{-1}, a difference of 19 kg ha^{-1}. However, the amounts of nitrogen that were taken up by the crops were almost the same; i.e. 179 kg ha^{-1} and 185 kg ha^{-1}, respectively, a difference of only 6 kg ha^{-1}. The figures obtained by subtracting the total amount of leached nitrogen and nitrogen taken up by crops from the applied amount of nitrogen were 411 kg ha^{-1} for the former and 106 kg ha^{-1} for the latter case. This difference was considered to be caused by denitrification in the dry field.

Another experiment was conducted using a lysimeter. The amounts of nitrate released from chemical fertilizer and from wastewater sludge were compared. The results showed that the release pattern differs between soil types. For river sand, with both chemical fertilizer and the sludge, nitrate concentration in permeated water quickly increased immediately after the fertilizers were applied. The concentrations then drastically decreased in the case of the chemical fertilizer while the decrease rate for the sludge was slow. For gray lowland soil, nitrate was constantly released by a certain amount of chemical fertilizer while the concentration for the sludge gradually increased. This experiment was conducted without any plantation. These results indicate a beneficial effect of organic fertilizer in preventing the release of nitrate.

As for the utilization of livestock manure for agricultural land, many researchers have focused on the maximum amount of manure that can be treated in a field. It is becoming important, however, that amounts of manure applied are appropriate in terms of environmental conservation.

4.3.2 Cultivation techniques

A crop-rotation method has been proposed as a cultivation techniques to prevent nitrate pollution. The rotation is made between vegetables requiring a large amount of fertilizer and general crops (such as wheat or potato) requiring less fertilizer, or beans that do not require any nitrogen fertilizer. In this rotation, nitrate accumulated in soil by vegetable cultivation is absorbed by the general crops or beans. The general crops and beans are called "cleaning crops".

4.3.3 Bio-remediation

Nitrogen used in dry fields undergoes oxidation to become nitrate, and moves downward into shallow aquifers. Groundwater with nitrate eventually flows out or is pumped out of the aquifer and may be used to irrigate paddy fields. Nitrate is then reduced to nitrogen gas in the paddy soil, which is in an anaerobic condition. The denisity of denitrifying bacteria in paddy soils is greatest at a depth of 0 to 5 cm from the surface, where denitrification capacity is highest. Since readily decomposable organic matter in the soil works as an electron donor in the denitrification reaction, the denitrification capacity is higher in the surface zone where organic matter is available. Furthermore, denitrification is facilitated further when harvest residues such as straw is applied as a carbon source. In pot experiment in which ^{15}N-labeled nitrate nitrogen and rice straw powder as carbon source, were added and maintained in a pooling condition, 84% of ^{15}N was denitrified in 23 days, while the denitrification rate in the pot without the addition of rice straw was only 0.4%.

4.3.4 Comprehensive measures from non-point agricultural sources

A summary of measures for general agricultural non-point sources is given as follows:

- Surface drainage from paddy fields should be minimized. Drainage should not be performed for ten days after fertilization.
- The initial fertilization should be spot or line fertilization, and paddling and leveling should be performed thoroughly in the case of complete-mix fertilization.
- Drain systems of paddy fields should allow repetitive use, and the planting period should be shifted from upstream to downstream.
- Conversion to crops with small fertilizer requirements should be conducted for dry fields, to prevent excessive fertilization. Slow-release fertilizers or effect-adjusted fertilizers should be combined efficiently to suppress eluviation of fertilizer components.
- Crop rotations including beans should be adopted as much as possible.
- Composting systems from domestic animals should be enhanced, and chemical fertilizer should be reduced when compost is used as fertilizer.
- Cultivation methods to suppress permeation for mulch cultivation in dry fields and grass cultivation in orchards, etc. and prevent soil erosion, should be adopted.
- Irrigation should be avoided for dry fields which require additional fertilization.
- Flat-land forests and vegetation on slopes should be preserved.
- The annual replanting area for forest lands and permanent grasslands should be limited to one-tenth or less of the land area used in the basin.

4.3.5 Water treatment technologies

If the nitrate (and nitrite) nitrogen level of source water exceeds 10 mgN L^{-1}, some treatment process must be provided before the distribution of drinking water. However, no simple method to remove nitrate has been developed. Some removal methods currently available are ion exchange, reverse osmosis, and electrodialysis (physical processes), and denitrification (a biological method using denitrifying microorganisms to convert nitrate to nitrogen gas).

Glossary

CNP:	Chloronitrophen, a herbicide for paddy fields.
DDVP:	Dichlorvos, A Organophosphate Pesticide.
Denitrification:	A biological nitrogen conversion process from nitrate to nitrogen gas caused by heterotrophic bacteria which uses nitrate as the oxidation agent.
Dioxin:	A name generally given to a class of super-toxic chemicals, the chlorinated dioxins and furans, formed as a by-product of the manufacture, molding, or burning of organic chemicals and plastics that contain chlorine.
Eutrophication:	A condition in an aquatic ecosystem where high nutrient concentrations stimulate blooms of algae.

Factor analysis:	A type of multivariate statistical analysis. Factor analysis is used to uncover the latent structure (dimensions) of a set of variables.
Green Revolution:	A term referring mainly to dramatic increases in cereal-grain yields in many developing countries beginning in the late 1960s, due largely to use of genetically improved varieties.
Methemoglobinemia:	Nitrate nitrogen in drinking water is transformed to nitrite by bacteria in the digestive system. Nitrite oxidizes iron in the hemoglobin of red blood cells to form methemoglobin. This lacks oxygen-carrying capacity, and the condition known as methemoglobinemia occurs.
Nitrification:	A biological nitrogen oxidation from ammonium to nitrate caused by autotrophic bacteria which uses ammonium as the energy source.
PCDD:	Poly-chlorinated dibenzo-dioxin.
PCDF:	Poly-chlorinated dibenzo-furan.
PCNB:	Pentachloronitrobenzene. A organochlorine fungicide used for seed and soil treatments at planting.
PCP:	Pentachlorophenol. The greatest use of PCP is as a wood preservative (fungicide). It was banned in 1987 for these and other uses, as well as for any over-the-counter sales.
Pesticides:	Substances and chemicals applied to kill or control animal pests, insects, fungi and nuisance weeds.
Photolysis:	Degradation of chemicals including pesticides by photochemical reaction.
Reverse osmosis:	A water purification technology which uses a semi-permeable membrane that blocks salts or other solutes.
Run-off:	The flux of pesticide discharged from agricultural fields.

Bibliography

Aizawa T., Magara Y. (1992): Behavior of pesticides in drinking water purification system. Water Malaysia '92. 8th. Aspac-IWSA regional water supply conference and exibition. Technical paper 3 10D2-1 - 10D2-9. [This examines how organic phosphate herbicides are oxidized to oxon-phosphates with increasing toxicity, by chlorination.]

DeZuane J. (1997). *Handbook of drinking water quality*, second edition. John Wiley & Sons, New York, USA. 575 pp. [This handbook is an essential volume for engineers, water supply and treatment personnel, environmental scientists, public health officials, or anyone responsible for assuring the safety of drinking water.]

Jørgensen S. E. and Johnsen I. (1981). *Principles of environmental science and technology*. Elsevier Scientific Publishing Company, Amsterdam, Netherlands. 516 pp. [The purpose of this book is to discover methods and principles being used in order to understand environmental processes and to use this knowledge to solve concrete environmental problems. Throughout the book, the application of methods and principles has been illustrated by examples of real environmental problems.]

Magara Y., Aizawa T., Matsumoto N. and Souna F. (1994): Degradation of pesticides by chlorination during water purification, IWAQ 17th Biennial International Conference, conference preprint book, pp311 – 320. [This examines pesticide degradation and by-products during chlorination process.]

Biographical Sketches

Yuichi Fushiwaki is a Senior Researcher at Kanagawa Prefectural Institute of Public Health, where he has been in office since 1999. He graduated from the Faculty of Pharmaceutical Sciences of Science University of Tokyo in 1974. Then, he worked for Kanagawa Prefecture and Kanagawa Environmental Research Center from 1974 and 1999, respectively. In the meantime, he obtained a Ph.D. in Engineering from Yokohama National University in 1994. His field of work includes evaluation of environmental toxicology in river water and atmosphere using bioassay, and the behavior and environmental pollution of pesticides.

He has given lectures for the Faculty of Engineering, Yokohama National University and Faculty of Environmental Science, Tsukuba University as a part-time lecturer since 1994 and 2002, respectively. He is also a visiting researcher of the National Institute for Environmental Studies.

Yasumoto Magara is Professor of Engineering at Hokkaido University, where he has been on faculty since 1997. He was admitted to Hokkaido University in 1960 and received the degree of Bachelor of Engineering in Sanitary Engineering in 1964 and Master of Engineering in 1966. After working for the same university for 4 years, he moved to the National Institute of Public Health in 1970. He served as the Director of the Institute since 1984 for Department of Sanitary Engineering, then Department of Water Supply Engineering. In the meantime, he also obtained a Ph.D. in Engineering from Hokkaido University in 1979 and was conferred an Honorary Doctoral Degree in Engineering from Chiangmai University in 1994. Since 1964, his research subjects have been in environmental engineering and have included advanced water purification for drinking water, control of hazardous chemicals in drinking water, planning and treatment of domestic waste including human excreta, management of ambient water quality, and mechanisms of biological wastewater treatment system performance. He has also been a member of governmental deliberation councils for several ministries and agencies including Ministry of Health and Welfare, Ministry of Education, Environmental Agency, and National Land Agency. He performs international activities with JICA (Japan International Cooperation Agency) and World Health Organization. As for academic fields, he plays a pivotal role in many associations and societies, and has been Chairman of Japan Society on Water Environment.

Professor Magara has written and edited books on analysis and assessment of drinking water. He has been the author or co-author of more than 100 research articles.

©*Encyclopedia of Life Support Systems* (EOLSS)

128

URBAN WATER POLLUTION

Katsuhiko Nakamuro
Professor of Environmental Health, Setsunan University, Osaka, Japan

Fumitoshi Sakazaki
Research associate of Environmental Health, Setsunan University, Osaka, Japan

Keywords: Sanitary sewage, Storm drainage, Runoff, Combined sewage system, Separated sewage system

Contents

1. Urban sewage
2. Sanitary sewage
3. Effects of rainfall on sewer
4. Storm drainage
Glossary
Bibliography
Biographical Sketches

Summary

Urban sewages consist of mainly domestic sewage and underground water. Qualities of urban sewages depend on the customs of the inhabitants and the character of the city. Sanitary sewage mainly arises from cooking, bath, laundry and feces. In areas where sewer for sanitary sewage is not serviced yet, unprocessed drainings are led into nearest marshes, lakes and rivers. Sewage hydrographs may be influenced radically by inflows of rainwater. Rainfall may cause serious hydraulic overloading of sewers and treatment facilities. Flow in a combined sewerage system during a storm often exceeds the capacities of sewers and treatment plants and discharged directly. Urban water pollution may be derived from non-point-sources which associated with general land runoff, road runoff including highway, construction and industrial runoffs, and hydraulic overloading of sewers and treatment facilities. Those pollutions from non-point-sources are more difficult to manage.

1. Urban Sewage

Qualities of sewages are quite different by their source and, consequently, treatment methods of them are also different. Because urban sewages consist of mainly domestic sewage and little of industrial wastewater and underground water, qualities of sewages depend on the customs of the inhabitants and the character of the city. Urban sewages contain 1,000~2,000ppm of solids and 70~80% of soluble substances. Organic matters in sewages are proteins, fats, carbohydrates, and their catabolite, which came from feces and sanitary sewages.

2. Sanitary Sewage

To prevent water contamination, mainly industrial wastewater has been treated, and application of strict regulations has improved quality of natural environment water. On the other hand, treatments of sanitary sewage have been limited to sewer service and sanitary sewages are not treated in areas where sewer is not serviced. Then, portion of pollution from sanitary sewage has increased yearly. Sanitary sewage occurs mainly from cooking, bath and laundry. Quantities and qualities of sanitary sewages depend on the habits of family, family constitution, season etc. About a half of biological oxygen demand (BOD) and suspended solids (SS) of sanitary sewage occur from cooking, while nitrogen and phosphorus from feces. In areas where sewer for sanitary sewage is not serviced yet, unprocessed drainings are led into the nearest sewer in service to reduce pollution in marshes lakes and rivers.

3. Effects of Rainfall on Sewer

Sewage hydrographs may be influenced radically by inflows of rainwater to the system. Connections of downspouts from roofs to sanitary sewers usually are forbidden by regulations; nevertheless, they are encountered often. Large numbers of those connections can produce sharp flow peaks during each rainfall and may contribute to serious hydraulic overloading of sewers and treatment facilities.

Recent practice has been based mostly on installing two "separate" sewerage systems in communities, one to carry sanitary wastes to the treatment plant, and the other to transport storm flow directly to discharge points, usually without treatment. However, many older communities still have "combined sewers," designed to carry both sanitary wastewaters and storm runoff in the same pipeline. Flow in a combined sewerage system during a storm may be many times that during dry weather, often exceeding the capacities of sewers and treatment plants and requiring the direct discharge of some mixed sanitary sewage and storm runoff to streams. The overflows are highly undesirable but continue to exist because of the great expense of converting "combined" systems into "separate" ones or building sewers and treatment plants large enough to handle the peak flows during storms.

4. Storm Drainage

4.1. Overview

Sources of potential pollutants add up several sources from which potential pollutants may enter watercourses. Sometimes water pollution problems may be caused principally by substances originating from one source but, more commonly, they are the result of accumulated contributions from several sources. It is important to recognize the several sources and know something about the types of materials that may be contributed by each.

Urban water pollution may be derived from non-point-sources, which are associated with general land runoff, road runoff including highway, construction and industrial runoffs, and hydraulic overloading of sewers and treatment facilities. Usually, flows from point-sources can be treated and controlled before discharge. On the other hand, non-point-sources are more difficult to manage and must be approached in different ways.

4.2. Pollutants in Rainwater and Runoff in Urban areas

Less generally recognized as a major source of potential pollutants is urban storm runoff, which can contribute large amounts of suspended solids, oxygen-demanding materials, metals, bacteria, nutrients, and many other potential pollutants. In fact, urban storm runoff sometimes may contribute potential pollutants in quantities even higher than those in untreated sanitary sewage from the same community. The sanitary wastes of several potential pollutants may have a significant impact on stream quality. Storm drainage from urban districts is especially polluted. Presently although treatments of industrial sewages are advanced, measures of urban storm drainage are late. That is a reason that urban storm drainage has become the principal primary factor of pollution of marshes lakes and rivers. It is difficult to grasp of actual condition and to control outflow of pollutants because of sudden increase of rainfalls.

In urban districts, waste gases from factories and automobiles are taken in to raindrops. In addition, atmospheric contaminants and particles of soot and dusts accumulated on roofs and roads are washed away by rainfall. Mn, Cu, Zn and Pb etc. are included in those pollutants. Sediments are more on roads made of older asphalt than roads made of concrete or newer asphalt. Runoff of the city is gathered once in street inlet connected to side grooves of roads, and when it exceeds the capacity of the street inlet, it flows into sewers. In street inlet, contaminants are concentrated due to evaporating of water at the time of fair weather, and dissolved organic matter and ammonium salt increase because of a repugnance condition in street inlets.

4.3. Control of Storm Drainage

The Administrations have tried to reduce loads of pollution of non-point-source on rainwater. Flows of rain first stage, which contain high level of contaminants, are led to storage facilities using storm sewer of city sewage services, and treated by sedimentation processing. Storage facilities like adjustment ponds are installed in underground of roads or somewhere. Runoff from rainwater is dispensed to sewer and treated at sewage disposal plants in time of night or fair weather when less sewage flows into plants. Screening process will effectively remove pollutants because more SS are contaminated in runoff than in sewage in the time of usual weather.

In order to reduce influx of runoff from rainwater to sewer, roads are paved with asphalts of more water permeability, and runoff are once stored in pavement or runoff permeate to the earth. In addition arranging gravel into bottom in street inlets enables runoff to permeate from the base to the underground. Cleaning sediments on roads is possible to decrease the load from road surfaces at the time of rainy weather. Cleaning of street inlets is effective. It is also effective to improve life habit with cooperation of inhabitants, such as stopping dumping of rubbish on roads, cleaning of road and side grooves around the residence, and use of non-phosphorus detergents.

Glossary

Combined sewage system:	Sewage and storm drain are discharged through the same conduit. When storm drain becomes 2-3 times more than sewage, the mixture of storm drain and sewage is discharged to a river or sea.
Domestic sewage:	It arises from daily life and is discharged from general houses, hotels, restaurants, offices and so on. It is consists of sanitary sewage and wastewater. It is distinguished from industrial wastes.
Street inlet:	It collects storm drain from houses and street gutters and leads through lateral sewer to sewer pipe.
Storm drain:	When the vapor in the air condenses to become rain, it is pure water. Along its path of fall, it dissolves particles and gasses in the air to contain many substrates. Solute in the rain is mostly sodium and chloride, and calcium, manganese, potassium and sulfide. The concentration of all solutes varies from 0.75 mg/l to 32 mg/l.
Separated sewage system:	Exterior sewer pipes are distinguished between for sewage and for storm drain. Sewages are introduced to sewage disposal plant and storm drains are discharged to a river or sea.

Bibliography

James C. Lamb (1985), Water quality and its control, John Wiley and Sons, Inc. New York, USA. [This book describes characters of waste waters and describes matters that require attention in drainage processing.]

Yasuhiko Wada (1990), Model analysis of pollution from non-point source, Gihodo Syuppan Inc., Tokyo, Japan. [This shows the detail character of and the measures against to pollution from non-point source.]

Gordon Maskew Fair and John Charles Geyer (1954), Water supply and waste-water disposal, John Willey and Sons, Inc., New York, USA. [This provides detailed information about the plumbing for waste water drainage.]

Editorial committee of 'Manual of water service and draining' (1973), Manual of water service and draining, Maruzen, Inc., Tokyo, Japan. [This describes water treatment and waste water processing.]

Biographical Sketches

Katsuhiko Nakamuro is Professor of Pharmaceutical Sciences at Setsunan University, where he has been at the present post since 1994. He graduated from Gifu Pharmaceutical University and received the Bachelor Degree and the Master Degree in Pharmaceutical Sciences in 1967 and in 1969, respectively. After graduation, he worked for Division of Environmental Chemistry, National Institute of Health Sciences, the Ministry of Health and Welfare of Japan from 1969 to 1984, participating in fundamental research for deciding standard of water quality and establishing measurement method for water quality. In the meantime, he also obtained the Ph.D. in Pharmaceuticals from Gifu Pharmaceutical University in 1977 on studies of environmental toxicology of sodium selenate. He worked for Office of National Environment Board in Thailand as a technical expert and transferred the water quality technology from 1983 to 1984.

Since 1984, he is with the Department of Environmental Health where his research topics cover development of risk-assessment for pollutants in water and evaluating toxicities of environmental chemicals including mutagens in river, disinfection byproducts by chlorination and ozonation, endocrine disrupting chemicals.

He has written many books on health risk of water contaminants. He has been the author or co-author of more than 150 research articles. He was awarded from Japan Water Works Association on his research of formation mechanism of trihalomethane during aqueous chlorination in 1983, and awarded from Japan Society on Water Environment on his research of effect of coexisting metals for trihalomethane formation during water chlorination in 1989. He has served as the official in Japan Society of Water Environment, the

Pharmaceutical Society of Japan, Japan Society for Biomedical Research on Trace Elements, Japan Ozone Association, and Japan Research Association for the Medical & Hygienic Use of Ozone.

Fumitoshi Sakazaki is Research Associate of Pharmaceutical Sciences at Setsunan University, where he has been at present post since 1998. He was graduated from Kyoto University and received the Bachelor Degree and the Master Degree in Pharmaceutical Sciences in 1996 and in 1998, respectively. Since 1998, he is installed at Department of Environmental Health where his research topics cover development of risk-assessment for pollutants in water and evaluating toxicities of environmental chemicals including endocrine disrupting chemicals.

INDUSTRIAL WATER POLLUTION

Katsuhiko Nakamuro
Professor of Environmental Health, Setsunan University, Osaka, Japan

Fumitoshi Sakazaki
Research associate of Environmental Health, Setsunan University, Osaka, Japan

Keywords: Industrial flow, BOD, hazardous contaminant

Contents

1. Industrial flows
2. Wastewaters
3. Examples of draining
Glossary
Bibliography
Biographical Sketches

Summary

Industrial operations in a community also may have a major influence on the shape of its wastewater hydrograph. Municipal and industrial wastewaters are especially important point sources of potential pollutants. They frequently contribute major quantities of BOD, suspended matter, bacteria, metals, and many specific chemicals potential of causing a wide variety of problems in watercourses or downstream uses. Qualities of wastewater depend on the industry type of the factories. Wastewater discharged from food industries contains high BOD. From paper industry SS and colored wastewater are discharged. Dyeing industry and pharmaceutical industry may release toxic organic matter. Draining from oil refining industry contains oil content, sulfides, ammonia and phenol etc. In plating industry, dilute or concentrated draining is discharged and both of them include acid, alkali, cyanide, chrome etc. Wastewaters from ships contain BOD and suspended solids might be potential of causing health risk of infectious diseases.

1. Industrial Flows

Industrial operations in a community also may have a major influence on the shape of its wastewater hydrograph. In some instances, large water users operate around the clock, perhaps even 7 days a week. Hydrographs in those systems usually are flatter because the large and continuous industrial flow makes variations in household and other uses have less relative effect on the total flow from the community. Conversely, discharges from large water users that operate for perhaps only 8 hr a day may have the opposite impact on the flow pattern, accentuating daytime peaks.

2. Wastewaters

Municipal and industrial wastewaters are especially important point sources of potential pollutants; in fact, they frequently are viewed by much of the public as being responsible for most or all water pollution problems. They frequently contribute major quantities of oxygen-demanding materials (BOD), suspended matter, bacteria, metals, and many specific chemicals capable of causing a wide variety of problems in watercourses or downstream uses.

Wastewaters from ships and pleasure boats have received special attention in recent years, sometimes with little justification and often without sound judgment. Sanitary wastes from vessels contain oxygen-demanding materials and suspended solids, usually in quantities too small to cause significant problems in the receiving stream, although on rare occasions they conceivably might be capable of causing difficulties in crowded harbors. The principal concern about these wastes has been the potential health risk of transmitting diseases through body contact with water receiving them, through drinking waters, or through the consumption of shellfish and other aquatic life. It has not yet been documented that waterborne diseases actually have been transmitted in this fashion, because of discharges from boats, although the possible route of transmission obviously is there.

Control systems for sanitary and oily discharges from vessels have been implemented at considerable cost. However, they offer special regulatory problems because the thousands of individual boats are widely scattered and frequently on the move, making adequate surveillance through inspection virtually impossible.

Other wastewaters that only recently have been recognized as important are the by-products originating in the treatment of water for municipal and industrial supplies. These waters, mostly ignored for many years, include sludge accumulations in settling basing and wastewater flows from backwashing sand filters. They contain most of the suspended matter and microorganisms in the intake water as well as the chemicals added to facilitate their removal. Some water treatment plants also discharge large quantities of dissolved solids from ion-exchange processes or from other treatment systems needed to produce the required final water quality. Today, discharges from water treatment plants are subject to the same regulatory controls as wastewaters from other industrial operations.

3. Examples of Draining

3.1. Draining Containing High-density Organic Compounds

Industrial classification	Industry type	Products	Quality of drainage		Displacement
Foodstuff	Meat processing	Sausage, ham, bacon, etc.	pH BOD COD SS T-N T-P	7 300-600 200-400 100-300 50-80 10-15	50-100
	Fishery food	Canned food, bottled food, etc.	pH BOD COD SS	7-8.5 200-2000 200-1800 150-1000	200-5,000

			T-N	100-200	
			T-P	30-80	
	Beer manufacturing	Beer	pH	8-11	5,000-10,000
			BOD	500-2000	
			COD	800-1200	
			SS	250-1000	
			T-N	30-50	
			T-P	5-15	
Chemical industry	Animals-and-plants oil-and-fats manufacture	Oil-and-fats of animals-and-plants	pH	4-9	100-2,000
			BOD	100-2,000	
			COD	100-1,500	
			SS	400-1,000	
			T-N	20-30	
			TP	40-80	
Pulp paper & paper products manufacturing industry	Pulp manufacturing industry	Kraft pulp	pH	7-9	150-300 m^3/t pulp
			BOD	300-700	
			COD	500-1,500	
			SS	40-80	
			T-N	110	
			T-P	5	
		Sulfite pulp	pH	3.5-4.5	150-500 m^3/t pulp
			BOD	300-500	
			COD	500-1,000	
			SS	50-300	
			T-N	100	
			T-P	3	
		Semi-chemical pulp	pH	3-7	100-150 m^3/t pulp
			BOD	500-2,000	
			COD	1,000-3,000	
			SS	200-600	
			T-N	70	
			T-P	2	

Table 1: Examples of draining containing high-density organic compounds

3.2. Draining Containing Low-density Organic Compounds

In alcoholic manufacturing industry, distillation facilities discharge highly condensed waste fluid. It used to be processed by methane fermentation or activated sludge method, but recently it is usually evaporated to be concentrated and made into fertilizer or fodder, or is burned. Factories using chemical synthesis method discharge phosphoric acid, which is used as catalyst.

At paper manufacture and pulp industry, SS and coloration by bark, chip and shaving of wood become problem when wet Barker is used for tearing off bark of raw wood. Cooking melting, in which lignin and hemicelluloses are dissolved and cellulose is extracted, discharges drainage containing sulfite and being highly colored and condensed. It is mainly concentrated and combusted. Bleaching draining includes oxides of lignin, reducing sugars, organic acids and colored substances etc. Because chlorine in bleach process is instead by ozone, trihalomethane is no more discharged. Draining from some process includes clay and starch.

Industrial classification	Industry type	Products	Quality of drainage		Displacement
Foodstuff	Dairy	Butter, cheese, ice cream, etc	pH BOD COD SS T-N T-P	6.5-11 50-350 50-200 70-150 30-40 5-8	1,000-6,000
	Spirit grain, flour milling	Spirit grain, flour milling, wheat flour, fodder, etc.	pH BOD SS	6-8 20-400 400-600	50-4,000
Textile industry	Filature industry	Silk	pH BOD COD SS T-N T-P	6-8 150-300 70-150 50-100 20-30 3-8	80-200
	Spinning industry	Cotton spinning, wool spinning, spun silk	pH BOD COD SS T-N T-P	3.5-9 150-400 200-300 60-800 20-140 10-30	100-1,000
Chemical industry	Organic products	Coal-tar product, dye, medicine inter-mediate, plastic, etc.	pH BOD COD SS T-N T-P	1-13 100 1,000 200-500 20-150 10-200 10-20	50-500

Table 2. Examples of draining containing low-density organic compounds

In dyeing industry, various dyes, aids and mordant are used depending upon materials. Because many factories concerned plural materials, displacement, components and density of drainage always fluctuate. Draining from processes of paste pulling out, refining, bleaching and finish have high BOD and COD. Clearing waste fluid is usually dilute but colored, and sometimes toxic.

Draining from each processes of oil refining industry contain oil content, sulfides, ammonia and phenol etc.

At pharmaceutical industries for medicine, animal drugs and agricultural chemicals, important issues are about pH, BOD, COD, SS and sometimes harmful matters. Besides draining from chemical synthesis process, draining from process in which medicines are extracted from culture solution become problem in respects of water quality and water displacement.

3.3. Draining Containing Organic and Toxic Compounds

Industrial classification	Industry type	Products	Quality of drainage	Displacement
Leather	Tanning	Leather, leather tanning	pH 7-12 BOD 80-2,500 COD 100-1,000 SS 50-3,000 T-N 250-350 T-P 10-20 Chrome	30-600
Iron manufacture	Blast furnace	Blast furnace pig iron	Gas liquor pH 9-9.5 BOD 3,000-4,000 COD 2,000-3,500 SS 50 T-N 800-1,000 T-P 20-50 Phenols 1,000-5,000 Thiocyanates 200-800 Cyanides 40 Sulfides 200-400 Ammonia 3,000-7,000	10% of dry distillation charcoal
			Gas scrubbing draining pH 9-9.5 BOD 150-800 Phenols 50-500 Cyanides 0-300 Ammonia 500-2,000	1.5-5 m^3/t of dry distillation charcoal

Table 3: Examples of draining containing organic and toxic compounds

The steel industry uses the mass water, but recently, reuse ratio has been improved by treating effluent from every process. Among draining, draining from pickling, rolling cooling, waste gas washing and wet dust collection etc. are polluted, but most draining such as from indirect cooling of furnace, roll and etc. are not polluted. Draining from coke oven contain cyanides, phenols and ammonia. Draining for chilling blast furnace and converter etc. contain SS made of ore, coke and iron etc., and processed by thickener, and it is reused. Draining from rolling process contain oil content which used for various purposes on surface of iron.

3.4. Draining Containing General Inorganic Compounds

Industrial classification	Industry type	Products	Quality of drainage		Displacement
Chemical industry	Inorganic manufactured products	Soda products, calcium carbide	pH BOD COD SS 00 T-N T-P	1-9 20 40 1,000-2,0 60-100 2-50	500-2,000
Ceramics, Stone and clay product	Glass	Glass	pH BOD SS	7-9 20-70 150-300	50-5,000
	Concrete manufacture	Concrete and cement	pH SS	9-14 150-500	100-300

Table 4: Examples of draining containing general inorganic compounds

3.5. Draining Containing Inorganic and Toxic Compounds

Industrial classification	Industry type	Products	Quality of drainage		Displacement
Metal product	Electroplate	Electroplate	pH CN Cr Cu, Cd, Zn, etc.	1-2 20-200 40-150	10-100
Ceramics, Stone and clay product	Glass	Optical glass, special glass	Pb, Cd, Se, etc.		

Table 5: Examples of draining containing inorganic and toxic compounds

In plating industry, dilute draining are discharged regularly from washing process and high density draining are discharged temporally when aged reagents are exchanged. Both of them include acid, alkali, cyanide, chrome etc. Because in this industry, preventing drops and spilling of liquids means reducing pollution as well as reducing expense, efforts to increase efficiency have been made. Methods free from harmful matters such as cyanides have been developed, but on the other hand, handling electronic parts increases new contaminants such as phosphoric acid and hydrofluoric acid.

Glossary

BOD:	Is abbreviation of biological oxygen demand. Organic matters in water need oxygen when they are degraded by microorganism aerobically. Bod indicates concentrations of organic contaminant of river, sewage, industrial wastes and so on.
Industrial Waste:	Wastewater discharged from factories or offices. Some

wastewater contains organic contaminant with high BOD, and others contain organic or inorganic hazardous contaminants.

Bibliography

James C. Lamb (1985), Water quality and its control, John Wiley and Sons, Inc. New York, USA. [This book describes characters of waste waters and describes matters that require attention in drainage processing.]

Editorial committee of 'Technologies and laws for preventing pollution' (supervised by Industrial science and technology policy and environment bureau, Ministry of economy, trade and industry of Japan) (1995), Technologies and laws for preventing pollution, Japan Environmental Management Association for Industry, Tokyo, Japan. [This shows BOD values from each industrial sewage.]

Editorial committee of 'Manual of water service and draining' (1973), Manual of water service and draining, Maruzen, Inc., Tokyo, Japan. [This describes water treatment and waste water processing.]

Biographical Sketches

Katsuhiko Nakamuro is Professor of Pharmaceutical Sciences at Setsunan University, where he has been at the present post since 1994. He graduated from Gifu Pharmaceutical University and received the Bachelor Degree and the Master Degree in Pharmaceutical Sciences in 1967 and in 1969, respectively. After graduation, he worked for Division of Environmental Chemistry, National Institute of Health Sciences, the Ministry of Health and Welfare of Japan from 1969 to 1984, participating in fundamental research for deciding standard of water quality and establishing measurement method for water quality. In the meantime, he also obtained the Ph.D. in Pharmaceuticals from Gifu Pharmaceutical University in 1977 on studies of environmental toxicology of sodium selenate. He worked for Office of National Environment Board in Thailand as a technical expert and transferred the water quality technology from 1983 to 1984.

Since 1984, he is with the Department of Environmental Health where his research topics cover development of risk-assessment for pollutants in water and evaluating toxicities of environmental chemicals including mutagens in river, disinfection byproducts by chlorination and ozonation, endocrine disrupting chemicals.

He has written many books on health risk of water contaminants. He has been the author or co-author of more than 150 research articles. He was awarded from Japan Water Works Association on his research of formation mechanism of trihalomethane during aqueous chlorination in 1983, and awarded from Japan Society on Water Environment on his research of effect of coexisting metals for trihalomethane formation during water chlorination in 1989. He has served as the official in Japan Society of Water Environment, the Pharmaceutical Society of Japan, Japan Society for Biomedical Research on Trace Elements, Japan Ozone Association, and Japan Research Association for the Medical & Hygienic Use of Ozone.

Fumitoshi Sakazaki is Research Associate of Pharmaceutical Sciences at Setsunan University, where he has been at present post since 1998. He was graduated from Kyoto University and received the Bachelor Degree and the Master Degree in Pharmaceutical Sciences in 1996 and in 1998, respectively. Since 1998, he is installed at Department of Environmental Health where his research topics cover development of risk-assessment for pollutants in water and evaluating toxicities of environmental chemicals including endocrine disrupting chemicals.

CONTAMINATION OF WATER RESOURCES

Yuhei Inamori
National Institute for Environmental Studies, Tsukuba, Japan

Naoshi Fujimoto
Faculty of Applied Bioscience, Tokyo University of Agriculture, Tokyo, Japan

Keywords: hazardous substance, eutrophication, hazardous microorganism, drinking water.

Contents

1. Introduction
2. Contamination by Hazardous Substances
3. Eutrophication
4. Contamination by Hazardous Microorganisms
Glossary
Bibliography
Biographical Sketches

Summary

Contamination of water resources by hazardous substances such as organochlorine compounds, agricultural chemicals and heavy metals harms living organisms and ecosystem. There is also the potential to harm human health through bioaccumulation through the food chain, contamination of public water supply, and recreational use of contaminated water resources.

Various problems such as decrease of dissolved oxygen, production of musty odors and toxic substances by cyanobacteria, and problems in waterworks are caused by algal blooms in eutrophic lakes and reservoirs. Throughout the world, many people suffer from water related diseases caused by cyanobacteria, pathogenic protozoa, pathogenic bacteria and pathogenic viruses.

1. Introduction

The environmental contamination of rivers, lakes, reservoirs, and ocean waters can be broadly categorized as contamination caused by organic chemicals, contamination caused by inorganic chemicals, contamination caused by microorganisms, and contamination related to physical conditions. Because these kinds of contamination threaten human health and living environments, the removal or degradation of contaminants and measures to control them are extremely important. This chapter discusses both chemicals and microorganisms causing contamination of water.

2. Contamination by Hazardous Substances

The growth and diversification of industry has been accompanied by wide use of diverse chemicals without adequate evaluation of their effects on the environment and on the human body, resulting in pollution of various kinds contaminating the environment. Examples include the contamination of soil and groundwater by improper treatment of hazardous substances, contamination of water bodies by inflowing wastewater containing hazardous substances, and contamination of soil, groundwater, and water bodies by the seepage of contaminants from landfill sites used to dispose of garbage. Contamination of water resources by hazardous substances harms living organisms and ecosystems at the same time as it harms human health through bioaccumulation, contamination of public water supply, and recreational use of contaminated water resources. The contamination of water and soil can hinder the growth of agricultural products and the contaminants can accumulate in the products. Contamination of lakes, reservoirs, rivers, and ocean waters can affect fish and shellfish. Contamination by hazardous substances seriously harms the agricultural and fishery industries in this way, and exposes people and animals to bioconcentrated contaminants. Table 1 shows a categorization of hazardous substances and typical substances of each kind.

Category	Typical substance
Organic compoundss	Dioxins, trichloethylene, PCB, benzene
Agricultural chemicals	Simazine, benthiocarb, thiuram
Heavy metals	Mercury, cadmium, lead, chromium, arsenic, selenium
Inorganic compounds	Nitrate, nitrite, cyanide

Table 1. Toxic substances

Agricultural chemicals is one class of hazardous substances that cause contamination; these are organic chemicals that include organochlorine compounds. They are extremely toxic and when humans are exposed to them, they can suffer from internal and neurological diseases. They include many substances that are suspected of being carcinogens or mutagens. Because people's health can also be harmed by exposure to these substances through drinking water, they are included in water quality standards for drinking water, environmental standards, and wastewater standards in many countries. In addition to lowland arable land, agricultural chemicals are spread in paddy fields, upland fields, in forests, on golf courses, in parks, and in green belts. They are also used in homes to fight insects. Because agricultural chemicals are generally highly soluble, they contaminate the environment by seeping into the ground with rainwater, while those on the ground's surface can be transported into river by rainwater. Contamination by organochlorine compounds attracted wide attention in the early 1970s. Typical organochlorine compounds are trichloroethylene (TCE) and tetrachloroethylene (PCE); these are used mainly as solvents or as dry cleaning agents. Many organochlorine compounds have been confirmed to be carcinogens through animal testing, and are suspected of acting as carcinogens on the human body.

Dioxins are now the best known of the organic chemicals. It is reported that 95% of dioxin is produced by incineration plants. Dioxin refers collectively to chemicals known as poly-chlorinated dibenzo-p-dioxin (PCDD) and poly-chlorinated dibenzofuran (PCDF). According to the substitution location and number of their chlorine atoms, there are 75 kinds of PCDD and 135 kinds of PCDF. The toxicity of these substances varies widely, with the most toxic being 2,3,7,8-TCDD, and the concentration of dioxins is often expressed as the total of values converted to the toxicity of 2,3,7,8 TCDD (Toxicity Equivalent: TEQ). As a result of investigation using laboratory animals, dioxins have been reported to be teratogenic, carcinogenic and immunotoxic, and to have toxic effects in reproduction . About half the dioxins discharged into the atmosphere fall to the ground with soot, where they contaminate the soil and water bodies. It is reported that about 90% of dioxin in people's bodies is ingested in their food, and dioxins in the soil and water bodies accumulate in people's bodies through vegetables, shellfish and fish. The WHO has established 4 pg $TEQ \cdot kg^{-1} \cdot day^{-1}$ as the quantity of dioxin that can be ingested without causing harm to people's health, even if they ingest it throughout their lives.

An agreement concerning persistant organic pollutants (Stockholm Convention on Persistant Organic Pollutants) was formally adopted in May 2001 at the United Nations Conference on Environment and Development. The agreement came into force in May 2004. Because Persistent Organic Pollutants (POPs) are chemical substances that resist decomposition in the environment (persistency), are easily concentrated in living organisms through the food chain (high accumulation property), travel long distances to readily accumulate in polar regions etc. (long distance commutation property), and harm human health and ecosystems (toxicity), and because it is reported that POPs cause global scale contamination (for example accumulating in the bodies of Inuit people, and in seals, camels, and other animals), action within an international framework was considered necessary. The substances covered by the Agreement are aldrin (insecticide), dieldrin (insecticide), endrin (insecticide), chlordane (insecticide), heptachlor (insecticide), toxaphene (insecticide), mirex (fire-retardant agent), hexachlorobenzene (insecticide), PCBs (insulating oils, heat transfer medium, etc.), DDT (insecticide), PCDD, PCDF, and fluorocarbons. Under this agreement, international cooperation programs are undertaken to provide technical and financial support to developing countries where measures to deal with these substances have not been implemented.

Inorganic chemicals that cause contamination include cyanide, nitrite, nitrate, and heavy metals such as cadmium, mercury, lead, and arsenic. Heavy metals and cyanide are extremely toxic, and many of these substances are acutely toxic and carcinogenic. In Japan, there have been two well-known cases of contamination by heavy metals. In one case, known as Minamata disease after the city where it occurred, organic mercury accumulated in fish through bioaccumulation, causing central nervous system disorders and fetal poisoning among the people who ate these fish. Another called *itai-itai* disease was an outbreak of osteomalacia and bone deformations among people who ate rice and drank water polluted by cadmium discharged by a mine. Contamination by nitrate and by nitrite in ground water is caused by excessive application of chemical fertilizers, inappropriate control of waste material from animal husbandry, and seepage of domestic wastewater into the ground. It is known that if infants ingest groundwater contaminated by a high concentration of nitrate nitrogen, they contract methaemoglobinaemia. Because the surfaces of soil particles are negatively charged, and negative ions such as nitrite and nitrate are not absorbed by the soil, they seep down to the aquifer, contaminating the groundwater. Another problem caused by nitrite nitrogen and nitrate nitrogen is that it forms N-nitroso compounds in people's stomachs. N-nitroso compounds has been proven to be carcinogenic to laboratory animals. Groundwater flows very slowly and has physical properties which mean that when it is influenced by heavy metals, volatile organochlorine compounds, or nitrate, dilution by further inflow does not normally occur. As a result, once such contamination has occurred, natural recovery is unlikely.

It has recently been discovered that there are chemicals that, when ingested by the bodies of living organisms, influence the hormonal activity that normally occurs in these living organisms. Such substances are called exogenous endocrine disruptors. It has also recently been reported that the mechanism of action of exogenous endocrine disruptors varies, with some acting directly through intranuclear receptors and others affecting the intracellular signal transmission system without acting through receptors. Examples of chemical substances that act through intranuclear receptors are those that bond with estrogen (female hormone) receptors to act in a way similar to the action of estrogen, and those that bond with androgen (male hormone) receptors to block the activity of androgen. Exogenous endocrine disruptors act in extremely minute quantities, accumulate in people's bodies, and are transmitted from mothers to their children, influencing later generations. They are not acutely toxic, and cause problems that appear after the children have grown up. The effects of these chemicals have also been reported in wildlife such as snails, carp, and alligators.

Selecting chemicals for inclusion in standards, and setting standard values for them, is done by (1) determining if each chemical substance actually effects human health, (2) evaluating the quantitative relationship of exposure to each chemical with its effects on health, (3) evaluating the degree to which people are exposed to each chemical, and (4) estimating the class and degree of risk to people and estimating the probability of health effects appearing in a specified population. The guideline values in the standards of the WHO have been calculated assuming that an adult weighing 60 kg drinks 2 liters of water a day, a child weighing 10 kg drinks 1 liter of water a day, and a baby weighing 5 kg drinks 0.75 liters of water a day. The guideline values are calculated accounting not only for the direct ingestion of chemical substances by drinking water, but also considering indirect ingestion such as the absorption of volatile substances and absorption through the skin during bathing and showering. Health risk assessments done to calculate the guidelines are based on data obtained by surveys of the effects of chemicals on people and on toxicity testing performed using experimental animals. This toxicity testing was performed to study acute exposure, short-term exposure, long-term exposure, reproductive toxicity, fetal toxicity, teratism, mutations, and carcinogenicity, etc. Then the NOAEL (no observed adverse effect level) and LOAEL (lowest observed adverse effect level) are obtained from data on each chemical substance. Depending on the chemical substance being considered, there are cases where the NOAEL and LOAEL are obtained using data from surveys of effects on people, and there are cases where they are obtained from toxicity experiments performed using experimental animals. The guidelines are established and calculated based on TDI (tolerable daily intake), i.e. the quantity of a chemical substance that will not endanger a person's health even if the person ingests it in food and water throughout his life.

3. Eutrophication

The eutrophication phenomena caused by the inflow of nitrogen and phosphorus is an extremely important form of contamination caused by inorganic chemical substances.

Water blooms occur, primarily in the summer, in eutrophic lakes and reservoirs such as Lake Kasumigaura, Tega Marsh, Inba Marsh, and Lake Suwa in Japan. The primary organisms that causes water bloom are cyanobacteria such as *Microcystis*, *Anabaena*, and *Oscillatoria*. Because these assimilate carbon dioxide, nitrogen, and phosphorus in the process of photosynthesis, they add to the load of organic material in the water, increasing COD and often leading to blooms and serious degradation of water quality. Because 1 mg of algae corresponds to 0.5 mg of COD_{Mn}, the increase of algae means the COD is rising. The annual average COD_{Mn} in Lake Kasumigaura in 1998 was 7.9 $mg \bullet l^{-1}$, i.e. much higher than the environmental standard value of 3 $mg \bullet l^{-1}$, but this high level is caused by the propagation of algae. As the water bloom that has propagated dies and is decomposed, the soluble organic carbon concentration rises and bacteria propagate, consuming oxygen. Therefore, the dissolved oxygen in the bottom layer declines after an algal bloom. This damages the habitats of fish and benthic organisms. In a temperate region such as Japan, water blooms occur every summer, interfering with water treatment processes by hampering coagulation, clogging sand filters, and causing odors, and adversely effect the fishing industry by reducing the oxygen needed by the bivalve *Corbicula* and nursery bred carp. In recent years, the filamentous cyanobacteria, *Phormidium tenue* and *Oscillatoria* have dominated, causing the musty odor of 2-methylisoborneol and geosmin. Deaths of animals that drank water containing water bloom have been reported in various countries. This occurs because the water contained toxins such as microcystin, produced by *Microcystis*, *Anabaena* and *Oscillatoria*. This toxin has claimed human lives in Brazil. In summary, the harmful effects of eutrophication Are as follows:

- Aggravation of water quality, decrease of dissolved oxygen in the bottom layer,
- Exposure to toxin produced by cyanobacteria through recreation and drinking water,
- Spoiling scenery, decrease of value for recreational use,
- Occurrence of bad smell,
- Clogging of sand filters, increase of costs due to addition of activated carbon treatment for removal of offensive odor and toxin

The growth of algae is influenced primarily by nutrients such as phosphorus and nitrogen, temperature and sunlight. In lakes and reservoirs with advanced eutrophication, not only cyanobacteria, but also chlorophyceae and bacillariophyceae appear, and the dominant species changes depending on the season. This occurs because the growth characteristics of different kinds of algae vary according to the nutrient concentration, the light intensity, and the temperature, with the result that the dominant species at a particular time is the one best adapted to the existing environment of each water body. It is reported that the total nitrogen concentration and total phosphorus concentration of lakes where the typical cyanobacteria, *Microcystis*, is dominant are more than 0.5 $mg \bullet l^{-1}$ and more than 0.08 $mg \bullet l^{-1}$ respectively. The authors performed a study of 211 lakes in Japan to find out which factors determine the domination by cyanobacteria, with nitrogen concentration, phosphorus concentration, and the N/P ratio as the study parameters. They reported that cyanobacteria dominate in a large percentage of the lakes and reservoirs where the total nitrogen and total phosphorus are more than 0.39 $mg \bullet l^{-1}$ and more than 0.035 $mg \bullet l^{-1}$ respectively. In lakes and reservoirs, the forms of nitrogen that can be utilized by algae are $NO3^-$, $NH4^+$ and , and the form of phosphorus that algae can use is $PO4^{3-}$. Nutrients are supplied to algae by various means, such as inflowing rivers, evacuation by zooplankton, the deamination reaction of organic nitrogen by bacteria, elution from sediments, rainfall, etc. As algae in lakes and reservoirs are supplied by nutrients from various sources, their growth rate is influenced by fluctuations in their concentration. Predators and decomposers are influenced by eutrophication, resulting in changes of biota to organisms that can exist under eutrophic conditions. If a breeding area in a lake or reservoir becomes anoxic as a result of eutrophication, it obviously lost as a spawning ground; although mature fish can withstand such external effects, their fry cannot. It is absolutely essential to implement measures such as nitrogen and phosphorus removal, as elevated concentrations have a great impact on the ecosystem.

It has been pointed out that it is important to control domestic wastewater; this nomally accounts for 40% to 70% of the total pollutant load on water bodies. The algal growth potential (AGP) test is a method of analyzing and evaluating the potential for treated water to encourage the growth of algae before domestic wastewater is discharged into a public water body. This test is done by sampling various kinds of wastewater and effluent in Erlenmeyer flasks, inoculating the water with algae, culturing it for between 10 and 14 days while irradiating it with light and controlling its temperature at about 20 ^{O}C until it grows to its maximum level, then measuring the quantity based on its dry weight and its chlorophyll A. Studying the growth potential in this way can clarify the degree to which the effluent can increase algal growth in the water where it is discharged. Table 2 shows the results of removal of nitrogen and phosphorus from domestic wastewater evaluated by the AGP test. It shows that the AGP result is reduced by the removal of both nitrogen and phosphorus. Such results were obtained for both algae that form the water bloom that appears in lakes and reservoirs and algae that forms red tides observed in inland seas. This AGP testing reveals that not only BOD, but phosphorus and nitrogen must be removed in sewerage treatment systems, wastewater treatment facilities in agricultural villages, and in small scale wastewater treatment plants.

	Effluent of wastewater treatment without N and P removal	Effluent of wastewater treatment with N removal	Effluent of wastewater treatment with N and P removal
T-N (mg/l)	31.0	7.5	6.5
T-P (mg/l)	3.8	3.5	0.1
AGP (mg/l)	420	80	10

Table 2. Effect of nitrogen and phosphorus removal on AGP test

4. Contamination by Hazardous Microorganisms

A United Nations survey has shown that about 1.4 billion of the world's 6 billion people cannot drink safe water, and that 20% of the world's population suffer from water-related diseases, and other health problems caused by a shortage of water needed to maintain hygienic conditions. It points out that pathogenic hazardous microorganisms originating in water include cyanobacteria such as *Microcystis, Anabaena, Oscillatoria, Aphanizomenon,* and *Cylindrospermopsis* that produce hepatotoxins and neurotoxins;

pathogenic protozoa such as *Cryptosporidium* and *Giardia*; pathogenic bacteria such as *Vibrio*, *Salmonella*, and *Shigella*; and more than 100 pathogenic viruses. Contamination by these pathogenic microorganisms is caused by inadequate water treatment in water supply and contamination in drainage systems. Because growth of algae is basically a result of eutrophication caused by inflowing phosphorus and nitrogen, growth of cyanobacteria could be held down by reducing the concentrations of phosphorus and nitrogen in the water body. Other hazardous microorganisms occur mainly by excrement contamination of water sources. It is possible to prevent infection by pathogenic bacteria and viruses by chlorine disinfection, etc., but pathogenic protozoa must be removed physically, because chlorine treatment alone cannot eliminate them.

A study to determine the quantities of these hazardous microorganisms that should be removed from water supply systems in order to supply safe water was carried out in America. Allowed concentrations of infectious organisms were obtained based on models of relationships between the quantity of hazardous organisms ingested and the probability of infection, based on the concept that it is possible to prevent infection if there is a risk of infection of one person from among 10 000 people drinking 2 liters of water from the public water supply every day for a year. For health assessment of hazardous chemical substances, the quantity ingested per 1 kg of body weight is important, but the important point when considering a hazardous microorganism that causes infection is whether or not it will be activated and grow inside the body of a person ingesting it. To prevent contamination of water sources by such infectious organisms, it is important to prevent excrement contamination, to provide more sewage treatment systems and small scale wastewater treatment plants, and to carry out thorough sterilization. But it is extremely difficult to completely eliminate hazardous microorganisms from water resources, because not only people but also wildlife can be infected by hazardous microorganisms. The water resources contain hazardous microorganisms because infected wild animals contaminate these waters with their excrement.

Contamination of water bodies includes contamination of physical properties of water such as color, as well as odor, and pH. Contamination of the physical properties of water does not directly harm people's bodies, but domestic use of the water can be very harmful. Problems related to the contamination of physical properties are color contamination, offensive odors, corrosiveness (pH), and turbidity. Color contaminations of water bodies is caused by metals, dye pollution, and soil particles, as well as by algal blooms caused by eutrophication. Offensive odor from water is caused by metals and by microorganisms that produce musty odors. Because residents are distressed if their drinking water and other domestic use water is colored or smells badly, water treatment must completely remove color and offensive odor from water. If the pH value of a water body is far from neutral, it effects the habitats of living organisms in the ecosystem, and high acidity of soil or agricultural water prevents the growth of plants and agricultural products. pH is, therefore, an important index of environmental pollution. Mechanical contamination occurs when exhaust gasses discharged by ships are dissolved by seawater, and when ships spill petroleum. Another cause of contamination is the ballast water that ships carry to remain balanced. International transportation and dispersion of hazardous marine microorganisms in ballast water has become an important problem. One aspect of this is the international movement of algal species that cause red tide, or algae that produce toxins, and although it is classified as biological contamination, essentially it can be viewed as physical contamination caused by the movement of seawater. A famous case of the effects of cross-border movement of ballast water is the harmful planktonic flagellate, *Alexandrium*, which was carried from Japan to Australia, where it caused serious problems by producing a toxin called saxitoxin. The *Alexandrium* has been designated as a biological weapon, and it is well known as an algal species whose international movement is strictly controlled. There are now several countries that strictly control the transport of ballast water from other regions, but in other countries, adequate protective measures have not been taken. Standards are now being established to develop practical technology so that ship owners can be required to biologically, physically, and chemically treat ballast water.

Glossary

AGP:	Algal Growth Potential
BOD:	Bio-chemical Oxygen Demand
COD:	Chemical Oxygen Demand
DDT:	Dichlorodiphenyltrichloroethane
LOAEL:	Lowest Observed Adverse Effect Level
NOAEL:	No Observed Adverse Effect Level
PCB:	polychlorinated biphenyl
PCDD:	Poly-chlorinated dibenzo-p-dioxin
PCDF:	Poly-chlorinated dibenzofuran
PCE:	Tetrachloroethylene
POPs:	Persistent Organic Pollutants
TCE:	Trichloroethylene
TDI:	Tolerable Daily Intake
TEQ:	Toxicity Equivalence
WHO:	World Health Organization

Bibliography

Okamoto H. (2000). Dioxin osen, Taikiken no kankyou (ed. Arita M.), 121-127, Tokyo Denki University Press, in Japanese. [This introduces toxicity and emission source of dioxins.]

Fujimoto N., Fukushima T., Inamori Y. and Sudo R. (1995). Analytical evaluation of relationship between dominance of cyanobacteria and aquatic environmental factors in Japanese lakes, Journal of Japan Society on Water Environment, 18, 901-908, in Japanese. [This paper investigates effect of nitrogen concentration, phosphorus phosphorus concentration, N:P ratio, water depth and mixing on dominance of cyanobacteria based on the data of Japanese lakes.]

Hirata T. (1998). *Cryptosporidium ,Giardia* tou no genchuu niyoru suidousui no risk no jittai, Mizu no risk management jitsumushishin (ed. Tsuchiya E, Nakamuro K. and Sakai Y.), 99-103, Science Forum, Tokyo, in Japanese. [This work introduce risk of infection of pathogenic protozoa *Cryptosporidium* and *Giardia* caused by tap water.]

Inamori Y., Fujimoto N. and Kong H. N. (1997). Recent aspect and future trend of countermeasures for renovation of nitrate-contaminated groundwater, Journal of Water and Waste, 39, 936-944, in Japanese. [This work introduces present state of nitrate contamination and countermeasure.]

Morita M. (1998). Kankyousuichuu no dioxin no risk hyouka, Mizu no risk management jitsumushishin (ed. Tsuchiya E, Nakamuro K. and Sakai Y.), 385-389, Science Forum, Tokyo, in Japanese. [This work introduces toxicity of dioxins and risk assessment of dioxins in water environment.]

Watanabe M. F., Harada K. and Fujiki H. (1994). Waterbloom of blue-green algae and their toxins,

University of Tokyo Press, in Japanese. [This work introduces taxonomy, and biology of blue-green algae, production of toxin by blue-green algae, and its toxicity, structure analysis and quantitative analysis.]

Biographical Sketches

Yuhei Inamori is an executive researcher of the National Institute for Environmental Studies (NIES), where he has been at his present post since 1990. He received B.S. and M.S. degrees from Kagoshima University, Kagoshima Japan, in 1971 and 1973 respectively. He received a Ph. D. degree from Tohoku University, Miyagi, Japan in 1979.

From 1973 to 1979, he was researcher for Meidensha Corp. From 1980 to 1984, he was researcher of NIES. From 1985 to 1990, he was senior researcher of NIES. His fields of specification are microbiology, biotechnology and ecological engineering. His current research interests include renovation of aquatic environments using bioengineering and ecoengineering applicable to developing countries. He received an Award of Excellent Paper fom Japan Sewage Works Association in 1987, Award of Excellent Publication from the Journal "MIZU"in 1991, and Award of Excellent Paper from Japanese Society of Water Treatment Biology in 1998.

Naoshi Fujimoto is assistant professor of Tokyo University of Agriculture, where he has been at hispresent post since 2000. He received B.E. and M.E. degrees from Tohoku University, Miyagi Japan, in 1991 and 1993 respectively. He received a D.E. degree in civil engineering from Tohoku University in 1996. From 1996 to 1999, he was research associate of Tokyo University of Agriculture. His field of specification is environmental engineering. His current research interests include control of cyanobacteria and its toxins in lakes and reservoirs.

ORGANICAL CHEMICALS AS CONTAMINANTS OF WATER BODIES AND DRINKING WATER

Yoshiteru Tsuchiya

Department of Applied Chemistry, Faculty of Engineering, Kogakuin University ,Japan

Keywords: Cyanobacteria, blue-green algae, *Streptomyces sp.*, eutrophication, geosmin, 2-methylisoborneol, *Oscillatoria geminata, O. splendida, O. agardhii, Phormidium tenue, Anabaena macrospora, A. flos-aquae*, earthy musty odor, water-bloom, *Aphanizomenon flos-aquae*, anatoxin-a, anatoxin-a(s), aphantoxin, *Microcystis aeruginosa, M. viridis*, microcystin, noduralin, *Nodularia spumigena*, cylindrospermopsin, *Cylindrospermopsis raciborskii, Umezakia natans*, volatile organic compounds (VOCs), PVC, DEHA,DEHP, EDTA, NTA, pesticides, DDT, BHC, organotin, dioxins, PCDD, PCDF, Co-PCB, PFAs, PFACs, PFOA, PFOS, POPs, PPCPs, TDI, TEQ, PAHs, endocrine disrupter, disinfectant by-products, THMs, MX, TOX, mutagenicity, water area, WHO Guidelines for Drinking-Water Quality.

Contents

1. Introduction
2. Contamination of metabolites produced by aquatic microorganisms
3. Contamination by industrial chemicals
4. Pesticides
5. Unintentionally generate substances
6. Miscellaneous organic substances
7. Disinfectant by-products
8. Characteristics of organic pollutants in each water area
9. Emergent chemicals
Glossary
Bibliography
Biographical Sketch

Summary

Pollution by organic chemicals in the aquatic environment occurs by various mechanisms. Naturally occurring organic chemicals produced by aquatic microorganisms, such as 2-methylisoborneol and geosmin, which have an earthy-musty odor, and microcystin, which shows hepatotoxicity, contaminate surface water such as rivers, lakes and reservoirs. Industrial waste containing artificial chemicals flows into and contaminates many water areas. Volatile organic compounds (VOCs), pesticides, phenolic compounds, phthalates, and nitrogen-containing compounds, are often detected in polluted water. VOCs such as trichloroethylene, tetrachloroethylene and 1,1,1-trichloroethane, are most often found in groundwater. Dioxins and polynuclear aromatic hydrocarbons (PAHs), produced during combustion of organic materials, are also found in surface water. On the other hand, chlorination by-products such as trihalomethanes (THMs), haloacetic acids and MX, which are produced during the water purification process by reaction of chlorine and organic materials, are found in drinking water. In the WHO Guidelines for Drinking-Water Quality , levels are set for 28 organic constituents (i.e microcystin-LR, chlorinated alkanes, chlorinated ethenes, aromatic hydrocarbons, chlorinated benzenes and miscellaneous), 33 pesticides, and 9 disinfectant by-products, due to their health effects on humans. . In recent years, occurrence of PPCPs(pharmaceutical and personal care products) and perfluoroalkyl acids(PFAs, i.e. PFOS and PFOA) in aquatic environment has been recognized as emerging issues in environmental chemistry.

1. Introduction

Pollution of organic chemicals in the aquatic environment occurs from natural products of aquatic microorganisms and artificial contaminants from industrial chemicals or human wastes. However, the effects of organic pollutants in rivers, lakes and ponds, reservoirs, groundwater and drinking water, differ for each specific contaminant. In general, organic chemical pollutants affecting human health are regulated by legal limits such as WHO Guidelines for Drinking-Water Quality, EPA National Primary Drinking Water Regulations, water quality standard in waterworks law, environmental standards and wastewater standards. Estimated organic contaminants are as follows:

- Organic chemicals produced by algae or Actinomycetes which growing in the natural aquatic environment.
- Organic pollutants of industrial chemicals, from industrial activity;
- Water contamination by pesticides from agricultural or other uses;
- Organic chemicals from unintentionally produced contaminants during combustion or burning of organic materials such as polynuclear aromatic hydrocarbons (PAHs);
- Pollutants in waste from the human environment such as domestic effluent, and
- Newly produced materials during the water purification process of drinking water.

This chapter describes organic compounds suspected of contaminating the aquatic environment.

2. Contamination of metabolites produced by aquatic microorganisms

Water pollution from organic chemicals occurs under various condition and from various causes. Naturally occurring organic chemicals produced by aquatic microorganisms such as blue-green algae (Cyanobacteria) or *Streptomycetes sp.* often cause water pollution. Such aquatic microorganisims can show remarkable increases in eutrophic water areas such as lakes, ponds, rivers, and reservoirs, and produce various metabolites. The eutrophication of water area occurs when nutrients such as N or P are increased by input of industrial effluent or domestic waste. Natural organic chemicals related to contamination of water are metabolites of blue-green algae and *Streptomycetes sp.* which contain volatile odorous compounds as well as toxic compounds.

2.1. Volatile metabolites

Volatile metabolites produced by blue-green algae and streptyomycetes are often found in water areas inhabited by these organisms. The volatile metabolites, 2-methylisoborneol (2-MIB) and geosmin (trans-1,10-dimethyl-trans-9-decalol), both of which produce an earthy/musty odor at an extremely low level (ng L^{-1}), may be contained in supplied drinking water. The problem of unpleasant drinking water has occurred in various countries such as USA, Canada, Brazil, Australia, Europe, South Africa, Japan and others. It is known that blue-green algae, *Oscillatoria geminata*, *Phormidium tenue* and others produce 2-MIB; *Oscillatoria splendida*, *Anabaena macrospora* and others produce geosmin, and *Streptomyces sp.* produce both compounds. They can cause a musty odor at concentrations in water of 5 to 10ng L^{-1}. These compounds are chemically stable without a double bond in the chemical structure as the third grade alcohol terpenoid.

For this reason, decomposition or removal of these compounds by water treatments such as chlorination or other common methods in water purification plants, is difficult. In Japan, aquatic microorganisms occur in about 40% of water sources every year in summer, and an earthy-musty odor develops in the water. From investigations in the 1970s, tap water with an earthy-musty odor is supplied every year to more than ten million people, and the odor is troublesome. Recently, however, ozone/activated carbon treatments have been carried out in many purification plants for water control, and the supply of water with an earthy-musty odor has decreased, to about one and a half million people in an investigation in 2000. Earthy-musty odors were investigated for one water area in Japan from April to November. 2-MIB was detected in the water supply source in the range of 2 to 1000 ng L^{-1} and geosmin at 1 to 500 ng L^{-1}. However, 2-MIB was detected at a maximum level of 12 ng L^{-1} and geosmin at 36ng L^{-1} in the tap water.

Some cyanobacteria may multiply even when the water temperature is low. For example, the highest rate of multiplication is shown by *Aphanizomenon flos-aquae* which inhabits ponds; a concentration of 7600 ng L^{-1} for geosmin was detected in the water when the water temperature was only 3.5 ^{O}C in winter. A comfortable guideline value for waterworks law in Japan recommends an extremely low concentration of 10-20 ng L^{-1} for 2-MIB and geosmin. These cyanobacteria also produce various metabolites such as aliphatic hydrocarbons (e.g. n-heptadecane, n-pentadecane, 1-heptadecene), fatty acids, amines (e.g.. trimethylamine), aromatic compounds (e.g. ethylbenzene), sesquiterpene alcohol and sulfur-containing compounds (e.g. methyl mercaptan, isopropyl disulfide).

2.2. Toxic substances produced by blue-green algae

Other species of blue-green algae containing toxic compounds produce blooms in eutrophic water bodies such as lakes, ponds and rivers. It has been found worldwide that certain animals (e.g. cow, horse, sheep, pig, chicken) can die after drinking water containing an algal bloom. The blue-green algae producing such toxic substances are as follows: *Anabaena flos-aquae*, *Aphanizomenon flos-aquae* produce anatoxin-a, anatoxin-a(s), and aphantoxin which shows neurotoxicity. *Microcystis aeruginosa*, *M. viridis*, and *Oscillatoria agardhii* produce microcystin which shows hepatotoxicity. The toxic substance, noduralin, is produced by *Nodularia spumigena,* and cylindrospermopsin is produced by *Cylindrospermopsis raciborskii* and *Umezakia* *natans*Microcystin has more than 50 isomers, but mirocystin–LR, –RR and –YR, are mostly found in water and in cells of algae. The typical chemical formula of microcystin is shown in Figure 1.

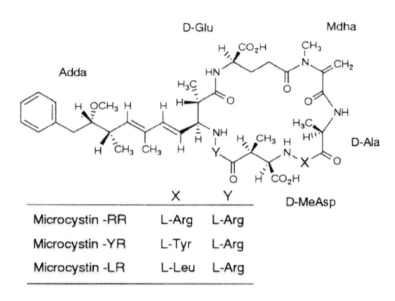

	X	Y
Microcystin -RR	L-Arg	L-Arg
Microcystin -YR	L-Tyr	L-Arg
Microcystin -LR	L-Leu	L-Arg

Figure 1. Chemical formula of microcystin

The first documented lethality to humans from cyanobacterial hepatotoxin (mycrocystin–LR), occurred via the intravenous route, in a dialysis clinic during 1996, when 76 deaths were reported. Microcystins are partly eluted to water by the environmental conditions, though they are generally contained in the cell body. Microcystins in the cells of cyanobacteria and in the water collected from water-blooms in eight water areas were analyzed. As a result, microcystin-RR was detected at a range of 0.03 to 8.8 μg L^{-1} in the water samples, and 0.15 to 205 μg L^{-1} in the algae body. Microcystin-YR was detected at 0.02 to 0.3 μg L^{-1} in the water sample and 0.09 to 9.7 μg L^{-1} in the algae body. Microcystin-LR was detected at 0.02 to 1.5 μg L^{-1} in the water sample and 0.07 to 390 μg L^{-1} in the algae body. About 1 to 5% of microcystins in the cell body are eluted to the water.

Microcystins in water decompose during reaction with chlorine, which is added for disinfection. For example, when 10 μg L^{-1}of each microcystin was treated with 1 mg L^{-1} of free chlorine, microcystin–RR and –LR decomposed in about 60 minutes, and microcystin–YR decomposed in about 20 minutes. However, chlorination by-products change based on the kind of substances involved in the interaction and the toxicity of these by-products is not clear. However, these three compounds showed a high degree of heat-resistance and could not be decomposed even by 30 minutes of boiling. In the WHO Guidelines for Drinking-Water Quality, microcystin-LR is recommended at 1 μg L^{-1} as a proposed guideline value. It is impotant to know whether the water supply source includes areas where there may be algal blooms.

3. Contamination by industrial chemicals

As a result of industrial activities, production of artificial chemicals and use of chemical products, various chemicals can flow into the environment and contaminate water bodies. Small amounts of organic chemicals are often detected in water. Among these contaminants, WHO guidelines for drinking-water quality has proposed recommended guidelines for limiting carcinogenic compounds and harmful compounds such as solvents, pesticides, raw materials from plastic products and organotin.

In Japan, about 2000 harmful substances were selected as priority research subjects. Field surveys of these chemicals in rivers, lakes, seas and their bottom sediments, and fish in the surrounding water area, have been carried out since 1974 by the Japanese Ministry of the Environment. The chemical substances were chosen according to the following priorities. Persistence in the environment, high bioaccumulation in the environment, chemical substances showing chronic toxicity, high consumption in the market. About 19% of the measured substances were detected in water as a result of investigating about 750 compounds at 10 to 400 sites between 1974 and 1998. Of the compounds detected in water, there were 22 volatile organic compounds (VOC), 6 phenolic compounds, 15 nitrogen-containing organic compounds, 5 pesticides, 5 PAHs (polynuclear aromatic hydrocarbons), 3 phthalates, 4 organotin compounds and 15 other substances. Of the VOCs detected, trichloroethylene, tetrachloroethylene and 1,1,1-trichloroethane were often found in water bodies at high concentrations. Other detected compounds were as follows: carbon tetrachloride, 1,2-dichloroethane, cis-1,2-dichloroethylene, trichloromethane, toluene, and others. Of the phenols, though bisphenol A which is an endocrine-disrupting chemical was not detected (0/60) (detected number/sample number) in 1976, it was detected in 27.7% (41/148) of water areas in the range of 0.01 to 0.268 ng L^{-1} in 1996. Phenol and p-butylbenzoic acid were also detected. Of the nitrogen-containing substances, amines and amides such as acrylamide, aniline, and cyclohexylamine were detected in river and lake water. Though aniline was detected at 58.5% (40/60) in the range of 0.02 to 28 µg L^{-1} by the field survey in 1976, it was only detected in one sample among 141, at 0.07 µg L^{-1}, during the investigation in 1998, indicating a clear decrease in water pollution. However, though 2-methylpyridine was not found (0/30) in 1986, an investigation in 1995 detected it at a range of 0.1 to 2.4 µg L^{-1} in 11.7% of samples. PAHs such as acenaphtylene, acenaphtene, benzo[a,h]anthracene, pyrene and benzo[ghi]perylene were detected in 1 to 4 samples at the 0.01 to 1 µg L^{-1} level.

As for other detected compounds in this field survey, ABS was detected in 17.6% of samples in the range of 280 to 29 000 µg L^{-1}, and phosphate substances (i.e. tributyl phosphate, trisbutoxyethyl phosphate, and tris-2-chloroethyl phosphate) were detected in 7.2 to 51.4% of samples in the range of 0.01 to 2.8 µg L^{-1}. It is clear that various chemical substances used cannot be detected in environment waters. As for organotin compounds , investigations have been carried out by the Ministry of the Environment since 1991. Though organotin compounds were used as agricultural agents, tributyltin(TBT) and triphenyltin(TPT) were specially used as antifouling agents to paint the bottoms of ships and fish nets and cages used in culture. Recently, organotin compounds have been shown to be endocrine-disrupting chemicals. The result of a field survey between 1991 and 2000 based on 100 samples of seawater from 37 sites collected from estuaries, ports and bays is shown in Table 1.

Tributyltin was detected in about 68% of samples in 1991, but decreased gradually thereafter, being detected in 8.8% of samples in 2000. Though the detected concentration was 0.08 µg L^{-1} in 1991, it decreased to a concentration of 0.005µg^{-1} in 2000. However, the detection frequency of triphenyltin was lower, in 6 to 10% of samples, on the initial survey, and hardly detected after 1995. The concentration of detected TPT showed 0.04 µg L^{-1} in the beginning, but has been less than 0.01 µgL^{-1} since 1995. In 1990, the Japanese government specified TBT as a special chemical compound, and restricted its use. It seems that this regulation has decreased the sea area affected by organotin compounds.

	Tributyltin(TBT)		Triphenyltin(TPT)	
	Detection rate (%)	Detection range µg L^{-1}	Detection rate (%)	Detection range µg L^{-1}
1991	60/93 (64.5%)	Nd-0.067	5/87 (5.7%)	Nd-0.014
1992	52/99 (52.5)	Nd-0.084	10/90(11.1)	Nd-0.044
1993	42/99 (42.4)	Nd-0.049	2/90 (2.2)	Nd-0.011
1994	35/99 (35.4)	Nd-0.03	4/92 (4.3)	Nd^0.01
1995	31/105(29.5)	Nd-0.042	0/87	
1996	27/105(25.7)	Nd-0.014	0/108	
1997	21/107(19.6)	Nd-0.009	0/108	
1998	20/76 (27.6)	Nd-0.008	4/102(3.9)	Nd-0.0015
1999	16/105(15.2)	0.003-0.01	3/105(2.9)	0.001-0.004
2000	9/102 (8.8)	0.003-0.005	0/102	

Table 1. Concentration of TBT and TPT in seawater

3.1. Volatile organic compounds (VOCs)

Volatile organic compounds (VOCs) were often found in the environment because of their high volatility; they diffuse into environment when solvents are used. The diffused chemicals in the environment contaminate both the atmosphere and water bodies. Many volatile organic compounds are found in various water areas, e.g. benzene, toluene, xylene, dichloroethane (used as solvents), trichloroethylene, tetrachloroethylene and 1,1,1-trichloroethane (used in dry cleaning) and carbon tetrachloride (used as a raw material for chlorofluorocarbons). Dichlorobenzene, trichlorobenzene styrene and vinyl chloride are found in the water after using the raw materials of industrial products. Trichloroethylene, tetrachloroethylene and 1,1,1-trichloroethane, used for metal washing and dry cleaning, were often the cause of groundwater contamination. Characteristics of these VOCs include poor solubility in water, heavy specific gravity, low surface tension, and low viscosity. Therefore, released VOC in the environment can readily percolate into groundwater. The Japanese Ministry of the Environment is monitoring 3500 to 4500 water samples targeting 11 compounds listed in the environmental standard of 1999. These 11 compounds were all detected in samples investigated in 1999, showing a 2.5 to 4.6% detection rate, exceeding the standard value for trichloroethylene, tetrachloroethylene and 1,1,1-trichloroethane. Fifteen samples of trichloroethylene demonstrated a maximal value of 0.99 mg L^{-1}, and 23 samples of tetrachloroethylene demonstrated a maximum value of 0.19 mg L^{-1}, exceeding the standard values. Other compounds exceeding the standard value were carbon tetrachloride, detected at a maximal value of 0.003 mg L^{-1}, 1,2-dichloroethane detected at a maximal value of 0.35mg L^{-1} and 1,1-dichloroethylene detected at a maximal value of 0.031 mg L^{-1}.

As for the results of the monitoring investigation of groundwater pollution, 46.7% of samples contained tetrachloroethylene (maximal value of 60 mg/L), 29% trichloroethylene (maximal value of 110 mg L^{-1}), and 20% cis-1,2-dichloroethylene (maximal value of 48 mg/L). VOCs contaminating groundwater were decomposed by microorganisms living in soil and water. Tetrachloroethylene decomposes into cis- and trans-dichloroethylene through trichloroethylene decomposition, and 1,1,1-trichloroethane decomposes into 1,1-dichloroethylene and 1,1-dichloroethane. For this reason, dichloroethylene or dichloroethane was sometime detected in the field survey of VOC polluted groundwater performed by the Japanese Ministry of the Environment, often at a low concentration. These compounds have guideline values for drinking water recommended by WHO, in view of their health effects on humans.

Compounds	Guideline value ($\mu g\ L^{-1}$)	Remarks
chlorinated alkanes		
carbon tetrachloride	2	
dichloromethane	20	
1.2-dichloroethane	30	for excess risk of 10^{-5}
1,1,1-trichloroethane	2000(p)	
chlorinated ethenes		
vinyl chloride	5	for excess risk of 10^{-5}
1,1-dichloroethene	30	
1,2-dichloroethene	50	
trichloroethene	70(p)	
tetrachloroethene	40	
aromatic hydrocarbons		
benzene	10	
toluene	700	
xylenes	500	
ethylbenzene	300	
styrene	20	
benzo(a)pyrene	0.7	for excess risk of 10^{-5}
chlorinated benzenes		
monochlorobenzene	300	
1,2-dichlorobenzene	1000	
1,4-dichlorobenzene	300	
trichlorobenzenes(total)	20	
epichlorohydrin	0.4(p)	

(p): Provisional guideline value

Table 2. WHO Guidelines for Drinking-Water Quality for VOC

3.2. Raw materials of plastics

Raw materials, additives, plasticizers and hardeners of plastics often contaminate water during the manufacturing process, or through the use of plastic products. The plasticizer of PVC (polyvinyl chloride), di (2-ethylhexyl) phthalate (DEHP) and di (2-ethylhexyl) adipate (DEHA) are detectable in waste water and surface water at ng to $\mu g\ L^{-1}$ levels.

The WHO Guidelines for Drinking-Water Quality recommend a value of 8 µg L^{-1} for DEHP, 80 µg L^{-1} for DEHA and a proposed value of 0.4 µg L^{-1} for epichlorohydrin. Bisphenol A, which is a raw material for polycarbonate plastics, and phthalates are suspected to be endocrine-disrupting chemicals. Monitoring the presence of these compounds in Japanese water area, bisphenol A, DEHP and DEHA were detected in 100% of wastewater samples and more than 60% of those from other areas such as rivers, lakes and ponds. WHO also recommended guideline values for other substances such as acrylamide of 0.5 µg/L, hexachlorobutadiene of 0.6 µg L^{-1}, EDTA of 200 µg L^{-1}, nitrilotriacetic acid (NTA) of 200 µg L^{-1} and tributyltin oxide of 2 µg L^{-1}.

4. Pesticides

Pesticides, such as insecticides, fungicides and herbicides, are used worldwide in agriculture to maintain a stable harvest, and on golf courses and urban areas to maintain a pleasant environment. Sprayed pesticides can diffuse into the atmosphere, water and soil. Pesticides in the atmosphere and soil flow into water bodies with rainwater and contaminate rivers, lakes and reservoirs that are often used as sources of water supply. A large quantities of organochlorine pesticides such as DDT, BHC and drin agents have previously been used in many countries. Use in developed countries has been stopped, however, because they are known to be harmful to humans and other animals.

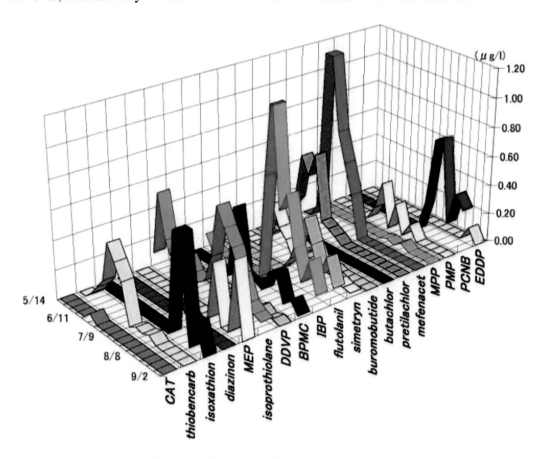

Figure 2. Pesticides found in river water

However, these agents are still extensively used elsewhere, both in agriculture and for vector control. DDT and BHC used in subtropical and tropical regions, are transferred to seawater within the Arctic circle by evapotranspiration, winds and precipitation. It is a serious concern that marine mammals inhabit the Arctic ocean and accumulate these pesticides in their fatty tissue. The pesticides detected in the water differ depending on the types of pesticides used, the period of use, and the purpose. For example, seasonal changes in pesticides in Japanese rivers are shown in Figure 2. In this study, 22 pesticides were detected in the river water, and herbicides used in paddy fields were highly prevalent in the water during June to July, during the weed growing period. When the source of water supply is contaminated with pesticides, tapwater is affected. Most pesticides could be removed or decomposed by water purification processes such as chlorination and ozonation, but chemically stable types of pesticides remain in the tapwater unless treated by a water purification plant. Figure 3 shows pesticides detected after treatment of river water. In this case, carbamate and phenoxy acid pesticides could not be remove by chlorination and was detected in tap water.

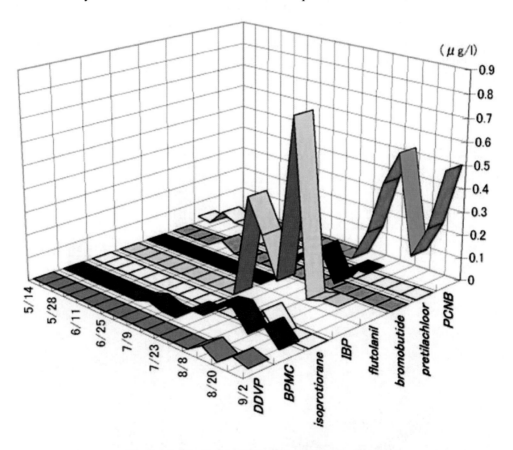

Figure 3. Pesticides found in treated water

Pesticides such as diazinon can be decomposed by chlorination, but the decomposed by-product sometimes shows toxicity. For example, after chlorination and ozonation, certain kinds of organophosphorous pesticides such as diazinon change into oxison derivatives which shows almost equivalent toxicity to diazinon Many types of pesticides are used worldwide, and various compounds detected in water depend on pesticide use. WHO Guidelines for Drinking-Water Quality recommend limits on 36 pesticides such as DDT, BHC, aldrin, and chlordane (see Table 3).

Compounds	Guideline value ($\mu g\ L^{-1}$)	Remarks
alachlor	20	for excess risk of 10^{-5}
aldicarb	10	
aldrin/dieldrin	0.03	
atrazine	2	
bentazone	300	
carbofuran	7	
chlordane	0.2	
chlorotoluron	30	
cyanazine	0.6	
DDT	2	
1,2-dibromo-3-chloropropane	1	for excess risk of 10^{-5}
1,2-dibromoethane	0.5-15(p)	for excess risk of 10^{-5}
2,4-D	30	
1,2-dichloropropane	40(p)	
1,3-dichloropropene	20	for excess risk of 10^{-5}
heptachlor/heptachlor epoxide	0.03	
hexachlorobenzene	1	for excess risk of 10^{-5}
isoproturon	9	
lindane	2	
MCPA	2	
methoxychlor	20	
metolachlor	10	
molinate	6	
pendimethalin	20	
pentachlorophenol	9(p)	for excess risk of 10^{-5}
permethrin	20	
propanil	20	
pyridate	100	
simazine	2	
trifluralin	20	
2,4-DB	90	
dichlorprop	100	
fenoprop	9	
mecoprop	10	
2,4,5-T	9	
diquat	10(p)	
terbuthylazine(TBA)	7	

(p): Provisional guideline value

Table 3. WHO Guidelines for Drinking-Water Quality for pesticides

5. Unintentionally generate substances

Among organic contaminants of water, by-products from incineration of waste or from combustion of gasoline also contaminate the environment and water bodies.

5.1. Dioxins

Polychlorinated-p-dibenzodioxin (PCDD) and polychlorinated-p-dibenzofuran (PCDF) are commonly called dioxins. They sometime involve coplanar PCB (Co-PCB) in the dioxins. Dioxins are mainly produced by incineration of waste, and they contaminate the atmosphere, water and soil. Dioxins are also produced by the metal refining industry, and contaminate their waste water. Dioxins are hard to decompose in the environment and easily dissolve into fat, particularly fish and shellfish living in contaminated water so that even though a very small quantity of dioxins can highly contaminate the environment through bio accumulation in the food chain. Tolerable daily intake (TDI) recommended by WHO is 1 to 4 pg TEQ kg^{-1} day^{-1}. In Japan, the recommended criterion for water quality in surface water is 1pg TEQ L^{-1}. About 3000 water samples including 1600 of river water, 100 lake water, 400 sea water and 1400 groundwater were analyzed in the year 2000 by the Ministry of the Environment in Japan. As a result, dioxins were detected at 0.014 to 48 pg-TEQ L^{-1} (average 0.36) in river water, 0.028 to 2.3 pg-TEQ L^{-1}(ave.0.22) in lake water, 0.012 to 2.2 pg-TEQ L^{-1}(ave.0.13) in sea water and 0.0008 to 0.89 pgTEQ L^{-1} (average 0.097) in groundwater.

5.2. Polycyclic aromatic hydrocarbons (PAHs)

PAHs have no industrial uses and are produced primarily as a result of the incomplete combustion of organic materials. The principal natural sources are forest fires and volcanic eruptions, while anthropogenic sources include the incomplete combustion of fossil fuels, wood stove emissions, coke oven emissions, aluminum smelters, and vehicle exhausts. PAHs are susceptible to aqueous photolysis under optimal conditions and they are biodegraded in water and taken up by aquatic organisms. Most of the PAHs in water are believed to result from urban runoff, from atmospheric fallout (smaller particles), and from asphalt abrasion (larger particles). The major source of PAHs varies, however, for each body of water. In general, most samples of surface water contain individual PAHs at levels of up to 50 ng L^{-1}, but highly polluted rivers have concentrations of up to 6000 ng L^{-1}. PAH levels in groundwater are within the range 0.02 to 1.8 ng L^{-1}, and drinking water samples contain concentrations of the same order of magnitude. Surveys of drinking water in Canada carried out in 1987 revealed benzo[k]fluoranthene (twice at 1 ng L^{-1}), fluoranthene, and pyrene. In 277 samples of municipal water supplies in the Atlantic provinces of Canada, taken during 1985 and 1986, fluoranthene, benzo[b]fluoranthene, benzo[k]fluoranthene, BaP, indeno[1,2,3-c,d]perylene and benzo[g.h.i]perylene were detected in 58, 4, 1, 2, 0.4 and 0.7% of samples, respectively. Major sources of PAHs in drinking water are asphalt-lined storage tanks and delivery pipes. The levels of individual PAH in rainwater ranged from 10 to 200 ng L^{-1}, whereas levels of up to 1000 ng L^{-1} have been detected in snow and fog. However, the main source of human exposure to PAHs is food, with drinking water contributing only minor amounts. WHO Guidelines for Drinking-Water Quality recommend 0.7 μg L^{-1} as the limit for benzo(a)pyrene.

6. Miscellaneous organic substances

Recently, numerous wildlife populations have been affected by the endocrine disruptive actions of various environmental contaminants. Suspected endocrine disrupting compounds are pesticides, PCB, dioxins, organotin, alkylphenols, phthalates, 2,4-dichlorophenol, benzophenone, 4-nitrotoluene. Monitoring the presence of these compounds in Japanese rivers, lakes, groundwater and sea water showed that PCB was detected in 77% of water areas at concentrations of nd to 150 ng L^{-1}. Nonylphenol, which is a decomposition product of the non-ionic detergent nonylphenol ethoxylate was found in more than 50% of water bodies, and benzophenone was detected in about 10% at a concentration of nd to 0.12 µg L^{-1}. Tributyl tin which is use as an antifouling agent was mostly found in seawater near the coast. However, the female hormone itself, 17 β-estradiol, was also found in 100% of waste water and more than 70% of water bodies at a concentration of nd to 0.021 µg L^{-1}.

7. Disinfectant by-products

Disinfection is unquestionably the most important step in the treatment of water for public supply. The destruction of microbiological pathogens is essential and almost invariably involves the use of reactive chemical agents such as chlorine and ozone, which are not only powerful biocides but also capable of reacting with other water constituents to form new compounds with potentially harmful long-term health effects. Thus, an overall assessment of the impact of disinfection on public health must consider not only the disinfectants but also their reaction products.

Disinfectant by-products are produced by the reaction of some organic chemicals in water with chlorine or ozone during chlorination or ozonation. The main organic substances that produce these by-products are humic materials such as humic acid and fluvic acid, algae bodies or their metabolites, and synthetic chemicals such as phenolic acid. Chlorination by-products such as trihalomethanes, haloacetic acid, haloacetonitrile, formaldehyde, chloral hydrate, MX and others are produced by reaction with chlorine. The mechanism involved in producing chlorination by-products involves a haloform reaction. In an investigation in the USA, about 40% of chemical structure in TOX (total organic halogen) is cleared, however, most substances have not yet been cleared. The distribution rate of the already known compounds in chlorination by-products are 69% trihalomethanes, 18% haloacetic acid, 9.0% haloacetonitrile, and 1.0% chloral hydrate.

7.1. Trihalomethanes (THMs)

Trihalomethanes (THMs) are halogen-substituted single-carbon compounds with the general formula CHX3, where X may be fluorine, chlorine, bromine, or iodine, or a combination thereof. With respect to drinking water contamination, only four members of the group are important: bromoform ($CHBr_3$), dibromochloromethane(DBCM)($CHBr_2Cl$), bromodichloromethane (BDCM)($CHBrCl_2$), and chloroform ($CHCl_3$). The most commonly occurring constituent is chloroform. Trihalomethanes occur in drinking water principally as products of the reaction of chlorine with naturally occurring organic substances and with bromide. The amount of each trihalomethane formed depends on the temperature, pH, and chlorine and bromide ion concentrations. This group of chemicals may act as an indicator of the presence of other chlorination by-products. Control of these four trihalomethanes should help to reduce levels of other uncharacterized chlorination by-products. These four compounds usually occur together and a number of countries have set guidelines or standards on this basis. WHO Guidelines for Drinking-Water Quality recommend 200 µg L^{-1} of chloroform, 60 µg L^{-1} of bromodichloromethane, and 100 µg L^{-1} of dibromochloromethane and bromoform. In a Canadian survey of the water supplies of 70 communities, the average concentrations of bromoform, DBCM, BDCM, and chloroform were 0.1, 0.4, 2.9 and 22.7 µg L^{-1}, respectively. THMs can be found in chlorinated swimming pools, total concentrations ranging from 120 to 660 µg L^{-1}.

7.2. Other chlorination by-products

MX (mutagenicity X), as 3-chloro-4-dichloromethyl-5-hydroxy-2(5H)-furanone, is formed by the reaction of chlorine with complex organic substances in water. It has been identified in chlorinated effluents of pulp mills and drinking water in Finland, UK, USA and Japan at concentration of 10 to 70 ng L^{-1}. MX is an extremely potent mutagen in strains of *Salmonella typhimurium* TA100, but the addition of liver extract S9 dramatically reduces the response. The contribution rate of MX for mutagenicity in tap water was about 10% to 50%. Haloacetic acids are synthesized by chlorination in water purification plants, and are frequently detected in tap water with trihalomethanes. Monochloroacetic acid at a range of 0.2 to 4.3 µg L^{-1} (ave. 1.3), dichloroacetic acid at a range of 1.3 to 9.4 µg L^{-1} (ave. 3.7), trichloroacetic acid at a range of 0.4 to 30.7 µg L^{-1} (ave. 6.1) and dibromoacetic acid at a range of 0.6 to 22.9 µg L^{-1} (ave. 2.6), were detected in tap water. Haloacetic acids synthesize the reaction of chlorine and ketones (e.g. acetone, acetylacetone, acetoacetic acid), aromatic hydrocarbons (e.g.. phenol, resorcinol) and nitrogen containing compounds (e.g. urasil, triethylamine and aspargic acid) by chlorination. It is reported that haloacetonitrile is detected in tap water as dichloroacetonitrile at a range of 0.2 to 13.0 µg L^{-1}, and trichloroacetonitrile at a range of 0.1 to 0.3 µg L^{-1}. Haloacetonitrile is synthesized as a reaction between chlorine and nitrogen-containing compounds such as amino acid. Dichloroacetonitrile is produced by chlorination of histidine, glutamine and phenylalanine, and dichloro- and trichloroacetonitrile which is produced by chlorination of fumic acid. Chloral hydrade is formed by dissolving trichloroacetoaldehyde in water, which is formed by the chlorination of the amino acid. Chloral hydrate is detected in tap water at a range of 1.0 to 10.3 (ave.4.2) µg L^{-1}.

7.3. Precursors of chlorination by-products

Disinfectant by-products are formed because of the organic materials (precursors) in water during the haloform reaction with chlorine. Various humic materials, algae bodies and their metabolites, and industrial chemicals act as precursors. Quality of humic materials consists of humic acid, fluvic acid and himatomeranic acid which appears mainly in the decomposition process of animals and plants. Though the chemical structure of humic acid is not clear, the molecular weight is estimated to be about 2500 to 10 000. It is clear that chlorination of humic acid and constituents of humic acid such as resorcinol, cathecol, hydroquinone and pyrogarol produce chloroform and other chlorination by-products. It was shown experimentally that the chloroform formation rate for humic acid is higher than that for fluvic acid.

As phytoplankton, there is a report that green algae such as *Scenedesmus sp.* and

Chlorella sp., cyanobacteria such as *Anabaena sp.* and *Oscillatoria sp.* inhabiting the water supply source can be isolated and pure cultures obtained. Then algae body and culture liquid medium solutions were separated and treated with chlorine. This experiment showed that the amount of chloroform increases as the cells increase, chloroform in metabolites is highly produced after a 50-day culture, and generally a higher amount of chloroform was obtained from algae body compared with that from culture metabolites. The amount of chloroform produced increased in the order of cyanobacteria, green algae and diatom. Among cyanobacteria, *Anabaena sp., Phorumidium sp.* and *Oscillatoria sp.,* which cause an earthy-musty odor, produce higher amounts of chloroform than *microcystis sp.,* which produce toxins such as microcystin. Trihalomethane (THM) formation potential in industrial chemicals is increased by aliphatic alcohols such as acetal, ketones such as acetone and acetylacetone. Aldehydes and carboxylic acid also show high production of acetone dicarboxylic acid while there is limited production from low molecular weight substances. Among aromatic substances, the substitution base is important to the THM formation potential. Acetophenone, which has an acetyl base, phenol which has a hydroxy group and aniline which has an amino base, all show high potentials.

In the case of amino acid, the amount of THM formation is higher with tyrosine and tryptophan.

8. Characteristics of organic pollutants in each water area

The characteristics of pollution with organic chemicals differ between each type of water body. Lakes and reservoirs used as sources of water supply are often densely populated with microorganisms and partially enclosed water areas often contain odorous metabolites or toxins due to these microorganisms. In rivers, industrial chemicals from industrial waste and small amounts of pesticides from agricultural use are mostly detected. In water treated with chlorine or ozone, disinfectant by-products such as trihalomethanes, haloacetic acid or decomposition products of pesticides are often detected. In well water, the volatile organic compounds trichloroethylene, tetrachloroethylene and their degradation products such as 1,1-dichloroethylene have been found. In seawater, organotin is in coastal areas, and seawater of the Arctic Ocean demonstrated small amounts of organochlorine substances such as PCB, DDT and HCH.

9. Emergent Chemicals

Recently, humans and veterinary used pharmaceuticals, perfluoroorganic chemicals, organobromine compounds such as PBDEs and HBCD which are flame retardant, cyanobacterial toxins, etc. are being detected as emergent chemicals from the environment. Especially, pharmaceuticals and perfluoroorganic chemicals are detected from the aquatic environment in many countries. It is noticed that these compounds effect or not to human health and ecosystem.

9.1 PPCP (Pharmaceuticals and Personal Care Products)

The presence and potential adverse effects of pharmaceuticals in the aquatic environment have been occurred to receive increasing attention in the popular and scientific press in recent years. Trace levels of pharmaceuticals detected in environmental samples, including sewage effluent, surface water, groundwater and drinking water in many countries such as Europe, USA, Canada, Australia, Japan and Korea. Detected pharmaceuticals in aquatic environment were analgesics and anti-inflammatroy drug such as diclofenac, ibuprofen, antibiotics (i.e. ciprofloxacin, clarithromycin, erythromycin-H_2O, sulfamethoxazole, trimethoprim) antiepileptic drug named carbamazepine, beta-blockers (atenolol, metoprolol, propranolol, sotalol), blood lipid regulators (i.e. clofibric acid, gemfibrozil benzafibrate), iodinated X-ray contrast media (i.e.iopromide, iopamidol, diatrizoate) and others. There are two main routes for pharmaceuticals to enter the aquatic environment. The first is via the effluent from sewage treatment plants (STP) after excretion from human body. Because, some pharmaceuticals such as carbamazepine and X-ray contrast media could not remove with present treatment systems. Removal efficiencies of pharmaceuticals in the STP are largely dependent on the facilities at individual STP. The second route is by the disposal of expired or unused pharmaceuticals. Main cause is via the disposal of sink and toilet, in the some case from household waste that is then taken to landfill sites. On the other hand, veterinary drugs may also cause pharmaceutical residues in the aquatic environment.

The pharmaceuticals in aquatic environment can consider the following effects for human or aquatic organisms, and pose a problem.

- Influence of the microbes on bacteria, a fungus, etc. It is generating of antibiotic resistance bacteria, especially.
- Influence of the ecosystems on an aquatic organisms etc.
- Influences on the human by contaminate to tap water by continuous exposure of two or more micro amount of pharmaceuticals. (*see also Water Supply and Health Care*)

9.2. Perfluorinated chemicals

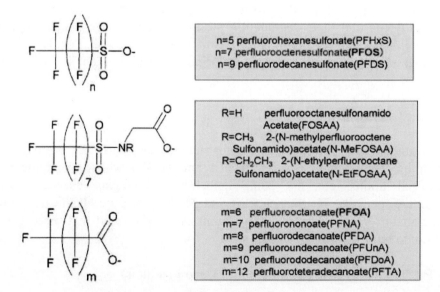

Figure 4. Chemical structure of PFACs

Polyfluoroalkyl chemicals (PFACs) have been used extensively since the 1950s in commercial applications, including surfactants, lubricants, paper and textile coatings, polishes, food packaging, and fire-retarding foams. Perfluorooctnesulfonate(PFOS) and perfluoroocanoate PFOA are members of a chemical group known as PFAs (perfluorinated acid), charcterised by chains of carbon atoms of varying lengths, to which fluorine atoms are strongly bonded. These are produced synthetically or by the end-stage metabolite of many PFAs. PFAs are heat stable, extremely resistant to degradation and environmental breakdown, and repel both water and oil. PFOS and PFOA have been found to be the dominant PFAs in the environment such as atmosphere, surface water, groundwater even drinking water, and in biota. In 2000, one of the fluorochemical manufacturing companies, the 3M Company, ceased the production of perfluorooctanyl-related materials because PFOS was found to be persistent in humans and wildlife. Chemical structure of PFACs is shown in Figure. 4

9.2.1. PFOS

PFOS (CAS chemical name: Perfluorooctane Sulfonate) is a fully florinated anion, which is commonly used as a salt or incorporated into larger polymers. The fluorine has the high electronegativity. The long chain alkyl group that fully substituted fluorine is chemical stable compounds for firm C-F binding. Due to its surface-active properties it is used in a wide variety of applications e.g. in textiles, and leather products; metal plating; food packaging fire fighting foams; floor polishes; denture cleansers; shampoos; coatings and coating additives; in the photographic and photolithographic industry; and in hydraulic fluids in the aviation industry.

PFOS is proposed as a candidate of POPs (Persistent Organic Pollutants) from Sweden.The reasons are as follows.

- Persistence: PFOS is extremely persistent. It has not showed any degradation in tests of hydrolysis, photolysis or biodegradation at any environmental condition tested. The only known condition wherby PFOS is degraded is through high temperature incineration.
- Bioaccumulation: PFOS fulfils the criteria for bioaccumulation based on the much higher concentration of PFOS in top predators, such as the polar bear, seal bald eagle and mink than at lower trophic levels.
- Potential for long-range environmental transport: PFOS meets the criteria for the potential for long-range atmospheric transport. This is evident through monitoring data showing highly elevated levels of PFOS in various parts of the northern hemisphere.
- Adverse effects: PFOS fulfils the criteria for toxicity. It has demonstrated toxicity towards mammals in sub-chronic repeated dose studies at low concentration (a few mg/kg bw/day) and displayed reproductive toxicity with mortality of pups occurring shortly after birth, probably caused by inhibition of lung maturation.

PFOS is toxic to aquatic organisms with the lowest NOEC(0.25mg/L) observed in mysid shrimp.

9.2.2. PFOA

PFOA (CAS chemical name: (Perfluorooctanoic acid) is also a fully florinated substance.

PFOA is chiefly used to help make high-performance materials known as fluoroelastomers and fluoropolymers. Fluoropolymers are high-performance plastic and synthetic rubber materials. They are used in harsh-chmical and high-temperature environments, primarily in performance critical applications in defense-related industries and in automotives, aerospace, electronics and telecommunications. Typical applications would be wire insulation for computer networks, semiconductor manufacturing equipment and automotive fuel hoses. About 95 percent of fluoropolymers are used in these types of industrial applications. The other 5 percent are used to make consumer products such as non-stick cookware and weather- and chemical-protective fabrics.

In February 2007, DuPont committed to no longer make, use or buy PFOA by 2015 or earlier, if possible. Taking this action because studies have shown very low levels of this compound in the environment and in the blood of the general population. Questions about this, as well as customer interest in product alternatives, are leading DuPont to develop new products and processes that reduce environmental footprint and are more environmentally sustainable.

9.2.3. PFOS and PFOA in aquatic environment

PFOS and PFOA, and other PFAs have been detected in aquatic environment such as surface water, groundwater and drinking water from many countries. By the case of the investigation in the global ocean water of PFOS and PFOA, it was detected in the ng L^{-1} order. Distribution map of PFOS and PFOA are shown in Figure 5.

PFOS and PFOA concentrations in the North Atlantic Ocean ranged from 8.6 to 36 pg L^{-1} and from 52 to 338 pg L^{-1}, respectively, whereas the corresponding concentrations in the Mid Atlantic Ocean were 13–73 pg L^{-1} and 67–439 pg L^{-1}. The surface waters of the North and Mid-Atlantic Ocean are more highly contaminated with PFOS and PFOA than are the surface waters of the South Pacific Ocean and the Indian Ocean (overall range of <5 to 11 pg L^{-1} for PFOS and PFOA). The highest concentrations of PFOS and PFOA found in the Atlantic Ocean were 15 and 88 times, respectively, higher than the lowest concentrations determined in the South Pacific Ocean and the Indian Ocean. The off-shore waters of the eastern North Pacific Ocean, along the Asia-Pacific Rim, contained 10–50-fold higher concentrations of PFOA than the water collected in the western North Pacific Ocean. Concentration of PFOA in the eastern Pacific Ocean ranged from 10 to 60 pg L^{-1} whereas that in the western Pacific Ocean ranged from 140 to 500 pg L^{-1}. Such a remarkable longitudinal (east–west) difference in the concentrations of PFOA in the Pacific Ocean was unexpected because many PFAs have been produced and used in large quantities in North America, thus the concentrations in the eastern North Pacific Ocean are expected to be high. Ocean water collected from the southern Indian Ocean contained trace levels of PFAs. PFAs were not found in the surface waters of the Antarctic Ocean, which is explicable by the isolation of the

Antarctic circumpolar water from the water masses in the Northern Hemisphere.

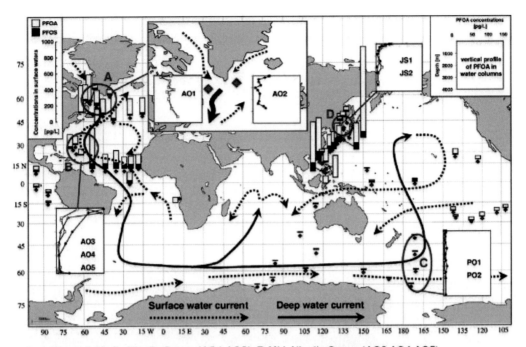

Locations; **A:**North Atlantic Ocean (AO1,AO2), **B:**Mid-Atlantic Ocean (AO3,AO4,AO5)
C: Central to Western Pacific Ocean (PO1, PO2), **D:** Japan Sea (JS1, JS2)

(Yamashita Chemosphere 2008)

Figure 5 Distribution map of PFOS and PFOA in ocean water

The vertical profile of PFAs in the two water columns from the Japan Sea (JS1 and JS2) showed gradual decrease in PFA concentrations from surface to subsurface layers (Fig. 5). Concentrations of PFOS in surface water samples were 15 pg L^{-1}, and decreased to $\sim$ 2 pg L^{-1} at a depth of 1000 m. At depths below 1500 m, concentrations were at or below the detection limit of 1 and 6 pg L^{-1} for PFOS and PFOA, respectively.The vertical profile of PFOS and PFOA concentrations in the Labrador Sea (AO1 and AO2) showed a remarkable difference from the JS1 and JS2 profiles from the Japan Sea (Fig. 5). Concentrations of PFOS and PFOA in the AO1 water column samples were relatively constant throughout the depth, except in subsurface water samples and water below 2000 m, in which PFOA concentrations increased. The AO2 water column samples from the Labrador Sea showed high concentrations of PFOS and PFOA at the surface and then uniform concentrations down to 2000 m. Temperature and salinity measurements for the AO1 and AO2 water columns suggested that the water mass from surface to a depth of 2000 m was well-mixed (Fig. 5). The Labrador Sea is known as the critical location for global circulation of ocean water and several investigations have revealed the role of this sea in moderating Earth's climate through its convective formation of the water mass known as Labrador Sea Water (LSW); This is the key element of the global thermocline circulation. Vertical profiles of PFAs in the marine water column were associated with the global ocean circulation theory.

The survey result of the New Jersey state on PFOS and PFOA in drinking water were described. Table 4. showed results of both compounds in the raw and finished drinking water.

Samples		PFOS ($\mu g\ L^{-1}$)	Detected number	PFOA ($\mu g\ L^{-1}$)	Detected number
Raw water	Well	0.0028 - 0.0049	7/15	0.0036 - 0.033	10/15
	Surface	0.0023 - 0.019	6/8	0.0036 - 0.035	8/8
Finished Water	Well	0.0038	1/3	0.0045, 0.0062	2/3
	Surface	0.0024 - 0.014	3/4	0.0030 - 0.027	4/4

Table 4. Concentration of PFOS and PFOA collected in New Jergey

PFOS was detected in 13 of 23 treatment plants or 57% of the public water system samples and PFOA was detected in 18 of 23 treatment plants or 78% of the public water system samples.Currently New Jersey recommends that the concentration of PFOA in drinking water be less than 0.04ppb.

9.2.4. PFACs in human and wildlife

PFACs have been found in the environment at the global scale. Therefore, it seems to accumulate it to wildlife and human. On the effect to the wildlife, many field surveys are carried out, and there is the many reports on a biological accumulations of PFACs. For example, 350ppb of PFOS and 8-12.9 ppb of PFOA are found from the sera of polar bears living in Greenland. About giant pandas in China, 11.2-12.7 ppb of PFOS and 0.56-0.90 ppb of PFOA, to albatrosses on the Midway Atoll, 16 ppb of PFOS and 180-182 ppb of PFOA are detected in the sera. The investigation of PFACs in human serum is carried out in many countries. The survey result of 12 countries was shown in Table 5.

Country	PFOS	PFHxS	PFOA	PFOSA
China	52.7	1.88	1.59	1.82
Japan	24.6	5.92	6.40	7.92
Korea	21.1	3.95	61.8	1.30
Malaysia	12.7	1.98	<10	4.57
Sri Lanka	5.03	0.57	6.38	na
India	1.85	1.60	2.64	<3
Belgium	15.7	1.04	4.82	<3
Poland	42.1	1.30	21.3	2.06
Italy	4.32	1.64	<3	1.78
Brazil	11.7	3.77	<20	1.00
Colombia	8.28	0.20	6.16	1.57
USA	43.2	4.00	19.1	4.14

Table 5. Perfluorochemical concentration in human blood sera

With these results, PFOS was detected from 1.85 to 52.7 ng mL^{-1}, and the China obtained highest value. PFOA was detected from <3 to 61.8 ng mL^{-1}, and it showed tendency low value than PFOS except Korean samples. Therefore, more investigations on PFACs in environment, and health effect to the human and wildlife are necessary.

Glossary

Anatoxin-a:	A toxin produced by *Anabaena flos-aquae*. It shows neurotoxicity.
Anatoxin-a(s):	A toxin produced by *Anabaena flos-aquae*. It shows neurotoxicity.
Aphantoxin:	A toxin produced by *Aphanizomenon flos-aquae*. It shows neurotoxicity.
BDCM:	Bromodichloromethane.
BHC:	Mixed isomers of 1,2,3,4,5,6-hexachloro-cyclohexane.
Co-PCB:	Coplanar polychlorinated biphenyl.
Cylindrospermopsin:	A toxin produced by *Cylindrospermopsis rachiborskii* and *Umezakia natans*. It shows hepatotoxicity.
DBCM:	Dibromochloromethane.
DDT:	1,1,1-trichloro-2,2-bis(4-chlorophenyl) ethane.
DEHA:	Di(2-ethylhexhyl)adipate.
DEHP:	Di(2-ethylhexhyl)phthalate.
Dioxins:	Polychlorinated-p-dibenzodioxin (PCDD) and polychlorinated-p-dibenzofuran (PCDF) are commonly called dioxins, and sometime involve coplanar PCB (co-PCB).
EDTA:	Etylene diamine tetra-acetic acid.
EPA:	Environmental Protection Agency of United State of America.
Geosmin:	trans-1,10-dimethyl-trans-9-decalol, a metabolite produced by cyanobacteria or *Streptomycetes sp.*
HCH	$(1\alpha,2\beta,\alpha3, 4\alpha,5\alpha,6\beta)$-1,2,3,4,5,6 hexachlorocyclohexane.
2-Methylisoborneol:	A metabolite produced by cyanobacteria or *Streptomycetes sp.*
2-MIB:	2-Methylisoborneol.
MX:	mutagenicity X, a chlorination by-product, 3-chloro-4(dichlorometyl) -5-hydroxy-2(5H)-franone.
Microcystin:	Toxins produced by the cyanobacteria *Microcystis sp.* It shows hepatotoxicity.
Noduralin:	A toxin produced by *Nodularia spumigena*. It shows hepatotoxicity.
NTA:	Nitrilotriacetic acid.
PAHs:	Polynuclear aromatic hydrocarbons.
PCDD:	Polychlorinated-p-dibenzodioxin.
PCDF:	Polychlorinated-p-dibenzofuran.
PFACs:	polyfluorinated chemicals
PFAs:	Perfluoroalkyl acids
PFOA:	Perfluorooctanoic acid
PFOS:	Perfluorooctanyl sulfonate
POPs:	Persistent organic pollutants
PPCPs:	Pharmaceuticals and personal care products
PVC:	Polyvinyl chloride.
TDI:	Tolerable daily intake

TEQ:	Toxicity equivalence.
THMs:	Trihalomethanes
TOX:	Total organic halide
VOCs:	Volatile organic compounds

Bibliography

Babcock D.B. and Singer P.C. (1979) Chlorination and coagulation of humic and fulvic acids, J.AWWA, **71,** 149-152. [This paper show production of trihalomethanes by humic acid and fulvic acid after chlorination]

Betts K. S. (2007) Perfluoroalkyl acids: What is the evidence telling us ? Environmental Health Perspectives, 115(5), A250-256 [This article shows a review of PFAs in environmet, wildlife and human]

Carmichael W.W, Azevedo S. et. al., (2001) Human fatalities from Cyanobacteria; Chemical and biological evidence for cyanotoxins, Envirn. Health Perspec. **109,** 663 –668. [This article shows human fatalities from cyanotoxins via the intravenous route]

Department of National Health and Welfare (Canada) (1997) national survey for halomethanes in drinking water, Ottawa.. [This report describes a survey of trihalomethanes in tap water at 70 water supplies]

Hoehn R.C Barnes D.B. et. al. (1980) Algae as sources of trihalomethane precursors, J.AWWA, **72,** 344 – 350. [This paper show production of trihalomethanes by algae body and their metabolites after chlorination]

Holmbom B., Voss R.H. et. al. (1984) Fractionation, Isolation, and characterization of Ames mutagenic compounds in Kraft chlorination effluents, Environ. Sci. Technol., **18,** 333-337. [This was the first article to identify MX in the chlorinated Kraft effluents and their strong mutagenicity]

Kümmerer K. (Editor) (2004) Pharmaceuticals in the environment: Sources, Fate, Effects, and Risks, Second Ed. Springer [This book shows PPCPs in the environment including sources, fate, effects and risks]

Matsushima R.N., Ohta T. et. al. (1992) Liver tumor promotion by the cynabacterial cyclic peptide toxin microcystin-LR, J Cancer Res Clin Oncol, **118,** 420-424. [This article show liver tumor promotion of microcystin-LR using animal experiments]

New Jersey Department of Environmental Protection, (2007) Determination of perfluorooctanoic acid (PFOA) in aqueous samples, Final Report [This paper described results of survey of PFOS and PFOA in the drinking water]

Suzuki N. and Nakanishi J. (1990) The determination of strong mutagen, 3-chloro-4-(dichloromethyl)-5-hydroxy-2(5H)-furanone in drinking water in Japan, Chemosphere, **21,** 387-392. [This article show production of MX in chlorinated drinking water in Japan]

Suzuki N. and Nakanishi J. (1995) Brominated analogues of MX in chlorinated drinking water, Chemosphere, **30,** 1557-1564 pp.[This article shows production of MX and brominated MX in chlorinated drinking water].

Ternes T. A. and Joss A. Edited (2008) Human pharmaceuticals, hormones and fragrances, IWA publishing[This book contains analytical methods, environmental risk assessment on PPCPS, also described waste water treatment, removal of PPCPs during drinking water treatmet]

Tsuchiya Y. and A.matsumoto (1988) Identification of volatile metabolites produced by blue-green algae, Wat. Sci. Technol. **20,** 149 pp. [This article identifies the volatile metabolites 2-MIB and geosmin from the 8 species of cyanobacteria isolated from a water body]

World Health Organization (1993). Guidelines for drinking-water quality. Second Ed. Volume 1, Recommendations.

World Health Organization (1996). Guidelines for drinking-water quality. Second Ed. Volume 2, Health criteria and other supporting information.

World Health Organization (1998). IPCS, Environmental health criteria monograph 205.Polychlorinated dibenzo-p-dioxins and dibenzofurans

World Health Organization (1998). IPCS, Environmental health criteria monograph 202. Selected non-heterocyclic polycyclic aromatic hydrocarbons.

Yamashita N. Taniyasu S. et. al. (2008) Perfluorinated acids as novel chemical tracers of global circulation of ocean waters (2008)-remove Chemosphere, **70** 1247-1255 [This article shows investigation of PFAs in open ocean water]

Yeung L. W. Y, So .M. K. et. al. (2006) Perfluorooctan sulfonate and related fluorochemicals in human blood samples from Chaina, E&ST, 715-720 [This article shows investigation of PFACs in human blood in China and other countries]

Biographical Sketch

Yoshiteru Tsuchiya is a Lecturer in the Faculty of Engineering of Kogakuin University, where he has been in his present post since 2000. He obtained a Bachelor Degree in Meiji Pharmaceutical College in 1964. He worked for Department of Environmental Health in Tokyo Metropolitan Research Laboratory of Public Health until 2000. In the meantime, he obtained a Ph.D in Pharmaceutical Sciences from Tokyo University. From 2002 to 2003, he worked for the Yokohama National University Cooperative Research and Development Center as Visiting Professor.

He has written and edited books on risk assessment and management of waters. He has been the author or co-author of approximately 70 research articles. He is a member of Japan Society on Water Environment, Pharmaceutical Society of Japan, Japan Society for Environmental Chemistry, Japan Society of Endocrine Disrupters Research, and International Water Association (IWA).

INORGANIC CHEMICALS INCLUDING RADIOACTIVE MATERIALS IN WATERBODIES

Yoshiteru Tsuchiya

Department of Applied Chemistry, Faculty of Engineering, Kogakuin University, Japan

Keywords: heavy metal, inorganic anion, radioactive materials, industrial waste, agricultural waste, domestic waste, drinking water, surface water, groundwater, arsenic, boron, fluoride, antimony, barium, cadmium, chromium, mercury nickel, cyanide, nitrogen, nitrate, nitrite, aluminum, copper, lead, gross alpha activity, gross beta activity, Itai-itai disease, Minamata disease, methyl mercury, methylation, methaemoglobinaemia, Alzheimer disease, Chernobyl, WHO drinking water guideline

Contents

1. Introduction
2. Naturally occurring substances in bodies of water
3. Inorganic substances in the industrial waste
4. Inorganic substances in agricultural and domestic waste
5. Inorganic substances in water supply
6. Radioactive materials
Glossary
Bibliography
Biographical Sketch

Summary

Most of the inorganic chemicals contained in aquatic environments are heavy metals, inorganic anions and radioactive materials. Therefore, the WHO guidelines for drinking water quality recommend values for 18 inorganic substances. Radioactive constituents are also recommended for gross alpha activity and gross beta activity.

Inorganic contamination of aquatic environments is caused by naturally occurring substances (fluoride, arsenic and boron), industrial waste (mercury, cadmium, chromium, cyanide and others), agricultural and domestic waste (nitrogen compounds), and systems for the distribution of drinking water (aluminum, copper, iron, lead and zinc). Naturally occurring inorganic materials mainly contaminate groundwater; industrial and agricultural waste, mainly surface water such as rivers, lakes and ponds; and pipes of distribution systems, mainly tap water.

Each of the inorganic substances contaminating water has a characteristic nature, concentration in water and effect on human health.

1. Introduction

Most inorganic chemicals polluting aquatic environments are heavy metals, inorganic anions and radioactive materials. This pollution is caused by naturally occurring substances such as fluoride, arsenic and boron, by industrial waste containing mercury, cadmium, chromium, cyanide and others, by agricultural and domestic waste containing nitrogen compounds, and from contamination by copper, iron, lead and zinc during the distribution of drinking water.

Naturally occurring inorganic substances mainly contaminate groundwater, whereas industrial and agricultural waste contaminates surface water such as rivers, lakes and ponds. The major source of contamination in drinking water is the distribution system. WHO guideline for drinking water quality recommended values for 18 inorganic substances. (Table 1)

Compounds	Guideline value (mg/L)	Remarks
Antimony	0.018	
Arsenic	0.01 (P)	For excess skin cancer risk of 6×10^{-4}
Barium	0.7	
Boron	0.01 (P)	
Cadmium	0.003	
Chromium	0.05	
Copper	2 (P)	
Cyanide	0.07	
Fluoride	1.5	*
Lead	0.01	**
Manganese	0.5	
Mercury(total)	0.001	
Molybdenum	0.07	
Nickel	0.02	
Nitrate(as NO_3^-)	50	***
Nitrite(as NO_2^-)	3	***
Selenium	0.01	
Uranium	0.009 (P)	

(P) : Provisional guideline value

* : Climatic conditions, volume of water consumed, and intake from other sources should be considered when setting national standards

** : It is recognized that not all water will meet the guideline value immediately; meanwhile, all other recommended measures to reduce the total exposure to lead should be implemented

*** : The sum of the ratio of the concentration of each to its respective guideline value should not exceed 1

Table 1: WHO drinking water guidelines inorganic constituents

Values are also recommended for radioactive constituents such as gross alpha activity (0.1 Bq L^{-1}) and gross beta activity (1 Bq L^{-1}).

This chapter describes the main substances contaminating water based on WHO guidelines.

2. Naturally Occurring Substances in Bodies of Water

Inorganic substances sometimes affect human health or ecological systems when eluted from the earth's crust into water. In many cases, materials such as arsenic, fluoride and boron are contained in groundwater; however, sometimes they flow to the surface of the earth as spring water.

2.1. Arsenic

Arsenic is a metalloid widely distributed in the earth's crust and present at an average concentration of 2 mg kg^{-1}. Arsenic can exist in four valency states; -3, 0, +3, and +5. Under reducing conditions, arsenite (As (III)) is the dominant form; arsenate (As (V)) is generally the stable form in oxygenated environments. Elemental arsenic is not soluble in water. Arsenic is used commercially and industrially as an alloying agent in the manufacture of transistors, lasers, and semiconductors, as well as in the processing of glass, pigments, textiles, and wood preservative.

Arsenic is also used in the hide tanning process and in pesticides, and pharmaceuticals. Arsenic is widely distributed in surface water, with concentrations in rivers and lakes generally below 10 µg L^{-1}. Arsenic levels in groundwater average about 1-2 µg L^{-1} except in areas with volcanic rock and sulfide mineral deposits where they can reach up to 3 mg L^{-1}. Concentrations of arsenic in the open ocean are typically 1-2µg L^{-1}. However, there have been a number of reports of higher than usual concentrations of arsenic in well water. Under natural conditions, the greatest range and highest concentrations of arsenic are found in groundwater as a result of the influence of rocks such as arsenopyrite (FeAsS). Chronic arsenic disease caused by arsenic-contaminated well water has been reported in various countries. Recent epidemiological study has revealed that a total of about 1 billion people; 47 million in India (West Bengal) and Bangladesh, 3 million in China, 0.3 million in Chile, and various numbers in other countries (i.e. Thailand, Nepal, Vietnam, Cambodia, Mexico, and Argentina) use arsenic-contaminated wells for drinking water, and this total is increasing year by year. However, it is not clear what the situation is in Africa, Central Asia and the Middle East because there has been no epidemiological study. Individuals developed chronic arsenic disease after drinking contaminated well water (about 0.5 mg L^{-1} As) for 5 to 6 years (average) or after drinking highly contaminated water (3- 5 mg L^{-1} As) for several months. Long-term exposure to arsenic in drinking water is causally related to an increased risk of cancer in the skin, as well as other effects on skin such as hyperkeratosis and changes in pigmentation. The Incidence of skin cancer is now being studied in Bangladesh, India, Chile and China.

Table 2 summarizes naturally occurring arsenic problems in groundwater reported from IPCS of WHO in 2001. In this report, arsenic was detected in groundwater at concentration of 176-5300 µg L^{-1}, and the population exposed ranged from 500 to 3 x 10^7. The WHO recommends 0.01mg L^{-1} for drinking water quality provisional guideline value. In a number of countries, this value has been adopted as a standard. However, many countries have retained the earlier WHO guideline of 0.05 mg L^{-1} as a national standard or as a target value.

Country/Region	Concentration Range (µg/L)	Area (km²)	Population exposed
Bangladesh	<0.5-2500	150,000	Ca. 3×10^7
West Bengal	<10-3200	23,000	6×10^6
Taiwan	10-1820	4,000	10^5
Inner Mongolia	<1-2400	4,300	?
China (Xinjiang)	40-750	38,000	500
Hungary	<2-176	110,000	29,000
Argentina	<1-5300	100,000	2×10^6
Northern Chile	100-1000	125,000	5×10^6
South west USA	<1-2600	5,000	3×10^5
Mexico	8-620	32,000	4×10^4

(IPCS ,EHC No.224 Arsenic 2001)

Table 2: Summary of naturally occurring As problems in groundwater

Countries that have adopted 0.01mg L^{-1} as the standard include members of EU(1998), Japan(1993), Jordan(1991), Laos(1999), Mongolia(1998) and Namibia, and Syria(1994). Countries that have adopted other values are as follows; Australia (0.007 mg L^{-1},1996), Canada(0.025mg L^{-1},1999), and USA(0.005 mg L^{-1}, 1986 proposed). Countries where the national standard for arsenic in drinking water remains at 0.05 mg L^{-1} include Bahrain, Bangladesh, Bolivia (1997), China, Egypt(1995), India, Indonesia(1990), Oman, Philippines(1978), Saudi Arabia, Sri Lanka(1983) Vietnam(1989) and Zimbabwe.

2.2. Boron

Boron is a naturally occurring element that is found in the form of borates in the ocean, sedimentary rocks, coal, shale, and some soils. It is widely distributed in nature, with a concentration of about 10 mg kg^{-1} in the Earth's crust (range: 5mg kg^{-1} in basalts to 100 mg kg^{-1} in shales) and about 4.5 mg L^{-1} in the ocean.

Boron is widely distributed in the environment, borax, kemite, and tourmaline being three of the more commonly mined boron minerals. The chemical forms of boron in nature include boric acid and more condensed species such as tetra-borate.

Elemental boron and its carbides are used in high-temperature abrasives, special-purpose alloys, and steel-making. Boron hydrides are used as reductants, to control heavy metal discharges in waste-water, as catalysts. Boric acid and borates are used in the manufacture of glass and as wood and leather preservatives and cosmetic products. Boric acid, borates, and perborates have been used as mild antiseptics or bacteriostats in eyewashes and mouthwashes. Borax is used extensively as a cleaning compound, and borates are applied as agricultural fertilizers.

Boron enters the environment mainly through the weathering of rocks, boric acid volatilization from seawater, and volcanic activity. Boron is also released from anthropogenic sources such as agricultural refuse, and from wood burning, the manufacture of glass, the use of borates/perborates in the home and industry, borate mining/processing, and sewage/sludge disposal.

Borate ions present in aqueous solutions are essentially in a fully oxidized state. No aerobic processes are likely to affect their speciation, and no biotransformation processes are reported.

The concentration of boron in seawater ranges from 4 to 5 mg L^{-1} as boric acid. Also naturally occurring boron is present in groundwater primarily as a result of leaching from rocks and soils containing borates and borosilicates. The concentration of boron in groundwater throughout the world ranges widely, from <0.3 to >100 mg L^{-1}. The amount of boron in surface water depends on the geochemical nature of the drainage area, proximity to marine coastal regions, and inputs from industrial and municipal effluents. Concentrations in surface water are summarized in Table 3.

Area	Boron concentration (mg/L)
USA	0.076 (median) , 0.387 (90th percentile)
Drainage basins	0.019-0.289
Coastal drainage waters	15 (boron-rich deposits)
lakes	157-360 (boron-rich deposits)
Canada, Ontario	0.029-0.086
river water	0.063
United Kingdom	0.046—0.822
Italy	0.4-1.0 (range of mean), <0.1-0.5
Sweden	0.013- (0.001-1.046)
Germany	0.02-2.0
Netherlands	0.09-0.145 (range of medians)
Austria, river waters	<0.02-0.6
Russia, river waters	0.01-0.02
Pakistan	<0.01-0.46 (near effluent discharges)
Turkey, river waters	<0.5 (uncontaminated), 4 (boron mine waste)
Argentina	<0.3, 6.9 (near borate plant)
Chile, river waters	3.99-26 (soil rich in minerals and natural salts)
Japan, river waters	0.009-0.012
South Africa	0.02-0.33

(IPCS,EHC No.204 Boron 1998)

Table 3: Concentration of boron in surface water

Concentrations in surface waters in North America (USA, Canada) ranged from 0.02 mg L^{-1} to as much as 360 mg L^{-1}. Concentrations ranged from 0.001 to 2 mg L^{-1} in Europe, with mean values typically below 0.6 mg L^{-1}. Similar concentrations have been reported for bodies of water within Pakistan, Russia, and Turkey; from <0.01 to 7 mg L^{-1}, with most values below 0.5 mg L^{-1}. In Japan, concentrations of boron in surface water ranged up to 0.1 mg L^{-1} and in South Africa up to 0.3 mgL^{-1}. Typical boron concentrations were less than 0.1 mg L^{-1}, with 90% of samples containing approximately 0.4 mg L^{-1}. In rain and snow, boron was found at 0.0031 to 0.0056 mg L^{-1}, respectively at six sites in western Switzerland.

Chronic exposure to boric acid and borax leads to gastrointestinal and kidney problems with loss of appetite, nausea, and vomiting, and the appearance of an erythematous rash. Boron is suspected to be an endocrine disrupter based on results of animal experiments in which it caused severe testicular atrophy and spermatogenic arrest.

2.3. Fluoride

Fluorine is a common element that does not occur in an elemental state in nature because of its high reactivity. It accounts for about 0.3 g kg^{-1} of the earth's crust and exists in the form of fluorides in a number of minerals, of which fluorite, cryolite, and fluorapatite are the most common. The oxidation state of the fluoride ion is -1. Inorganic fluorine compounds are used in industry for a wide range of purposes. They are used in aluminum production and as a flux in the steel and glass fiber industries. They can also be released into the environment during the production of phosphate fertilizers (which contain an average of 3.8% fluorine), bricks, tiles, and ceramics. For dental purposes, fluoride preparations may contain low concentrations of fluoride (1000-1500mg per kg of toothpaste). Fluoride levels in surface waters vary according to geographical location and proximity to emission sources. Surface water concentrations generally range from 0.01 to 0.3 mg L^{-1}. Ambient surface water fluoride levels are summarized in Table 4.

Location	Fluoride concentration (mg/L)
Areas with low natural fluoride	
Canada	0.05
USA	0.035-0.052 (<0.001-0.59)
Belgium, river waters	0.13-0.2
France, river waters	0.08-0.25
Norway, lakes waters	0.037 (<0.005-0.56)
Spain, river waters	0.1
United Kingdom, river waters	<0.05-0.4
streams	0.02-0.22
Nigeria, river waters	0.1-0.12
India, river waters	0.2-0.25
river waters	0.038-0.21
China, river waters	0.04
Japan,	>1
Areas with high natural fluoride	
USA、 hot springs	20-50
river waters	1-14
lakes	>13
Kenya, lakes	>2800

(IPCS, EHC No.227 Fluoride 2002)

Table 4: Concentration of fluoride in unpolluted water

Concentrations of fluoride in surface waters of North America ranged from 0.035 to 0.05 mg L^{-1}. Concentrations ranged from <0.05 to 0.56 mg L^{-1} in Europe, with similar concentrations reported for water bodies within Nigeria, India and China. Higher levels of fluoride have been measured in areas where the natural rock is rich in fluoride and near industrial outfalls. For example, in Norwegian lakes, concentrations ranged from <0.005 to 0.56 mg L^{-1}, with a mean value of 0.037 mg L^{-1}. Elevated inorganic fluoride levels in surface water are often seen in regions where there is geothermal or volcanic activity. The most well documented area of high-fluoride surface water associated with volcanic activity follows the East African Rift Valley system. Many of the lakes of the Rift Valley system have extremely high fluoride concentrations—for example, 1640 and 2800 mg L^{-1}, respectively, in the Kenyan lakes. In seawater, a total fluoride concentration of 1.3 mg L^{-1} has been reported.

Fluoride is probably an essential element for humans and animals. Fluoride is an important element and where there are elevated intakes through drinking water, sometimes in conjunction with other sources, crippling skeletal fluorosis, can occur in addition to the adverse cosmetic effects of dental fluorosis. Many epidemiological studies of possible adverse effects of the long-term ingestion of fluoride via drinking water have been carried out. These studies clearly establish that fluoride primarily produces effects on skeletal tissues (i.e. bones, teeth). Low concentrations provide protection against dental caries, especially in children. This protective effect, which is associated with surface contact with enamel, increases with concentration up to about 2 mg L^{-1} of drinking water; the minimum concentration of fluoride in drinking water required to produce it is approximately 0.5 mg L^{-1}. However, fluoride also can have an adverse effect on tooth enamel and may give rise to mild dental fluorosis at drinking water concentrations between 0.9 and 1.2 mg L^{-1}.

Elevated fluoride intakes can also have more serious effects on skeletal tissues. Skeletal fluorosis is observed when drinking water contains 3-6 mg L^{-1} of fluoride. Crippling skeletal fluorosis usually develops where drinking water contains over 10 mg L^{-1} of fluoride.

3. Inorganic Substances in Industrial Waste

Many inorganic substances are used for industrial manufacturing. Pollution of the water environment occurs from industrial manufacturing and the use of these products. Many kinds of metals contaminate water via waste water and sludge from industrial factories, mines or smelting stations. Metals related to the contamination of water are as follows; antimony, arsenic, barium, beryllium copper, iron, zinc, cadmium, chromium, manganese, nickel, selenium, mercury, uranium and others. Main inorganic substances related to water pollution and health effects are described.

3.1. Antimony

Antimony is used in semiconductor alloys, batteries, antifriction compounds, cable sheathing, flameproofing compounds, ceramics, glass, pottery, type castings for commercial printing, solder alloys and fireworks. Antimony has been identified in natural waters in both the antimony (III) and antimony (V) oxidation states and as methyl antimony compounds. It occurs in seawater at a concentration of about 0.2µg L^{-1}. A survey in the USA found antimony in only three of 988 samples of drinking water from groundwater sources, the concentrations ranging from 41 to 45µg L^{-1}. In a study of 3834 samples of drinking water, antimony was found in 16.5%, at concentrations ranging from 0.6 to 4 µg L^{-1}. Workers exposed for 9-31 years to dust containing a mixture of antimony trioxide and antimony pentoxide in an antimony smelting plant exhibited symptoms such as chronic coughing, bronchitis, emphysema, conjunctivitis, and staining of the teeth. Antimony dermatitis was seen in more than half of the exposed workers.

3.2. Barium

Barium is present as a trace element in both igneous and sedimentary rocks. Although it is not found free in nature, it occurs in a number of compounds, most commonly barium sulfate (barite) and barium carbonate (witherite). Barium compounds, including barium sulfate and barium carbonate, are used in the plastics, rubber, electronics and textiles industries, in ceramic glazes and enamels, in glass-making, brick-making, and paper-making, as a lubricant additive, in pharmaceuticals and cosmetics, and in the hardening of steel. Barium in water comes primarily from natural sources. The acetate, nitrate, and halides are soluble in water, but the carbonate, chromate, fluoride, oxalate, phosphate and sulfate are quite insoluble. The solubility of barium compounds increases as the pH level decreases.

The barium concentration measured in groundwater in the Netherlands was at 60 locations had a mean and maximum of 0.23 and 2.5 mg L^{-1}, respectively. In drinking water in Canada, the concentration was found to range from ND to 600 µg L^{-1} (mean 18µg L^{-1}). In a study of the water supplies of cities in the USA, a median value of 43µg L^{-1} was reported. Levels of barium in municipal water supplies in Sweden ranged from 1 to 20 µg L^{-1}.

Barium is not considered an essential element for human nutrition. The prevalence of dental caries was reported to be significantly lower in children from a community ingesting drinking water containing 8-10mg L^{-1} of barium as compared with that in children from another community ingesting drinking water containing <0.03 mg L^{-1}. Associations between the barium content of drinking water and mortality from cardiovascular disease have been observed in several epidemiological studies. Significant negative correlations between the barium concentration in drinking water and mortality from atherosclerotic heart disease and total cardiovascular disease have been reported.

3.3. Cadmium

Cadmium is a metal with an oxidation state of +2. It is chemically similar to zinc and occurs naturally with zinc and lead in sulfide ores. Cadmium metal is used mainly as an anticorrosive, electroplated on to steel. Cadmium sulfide and selenide are commonly used as pigments in plastics. Cadmium compounds are used in electric batteries, and electronic components.

The solubility of cadmium in water is influenced to a large degree by its acidity; suspended ore sediment-bound cadmium may dissolve when there is an increase in acidity. In natural waters, cadmium is found mainly in bottom sediments and suspended particles.

Cadmium concentrations in unpolluted natural water bodies are usually below 1 μg L^{-1}. Median concentrations of dissolved cadmium measured at 110 stations around the world were <1μg L^{-1}. Average levels in the Rhine and Danube river in 1988 were 0.1μg L^{-1}(range 0.02-0.3μg L^{-1}) and 0.025μg L^{-1}, respectively. Contamination of drinking water may occur as a result of the presence of cadmium as an impurity in the zinc of galvanized pipes or cadmium-containing solders in fittings, water heaters, water coolers, and taps. In the Netherlands, in a survey of 256 drinking water plants in 1982, cadmium (0.1-0.2μg L^{-1}) was detected in only 1% of the drinking water samples. In Japan, in a survey of 5550 drinking water plants in 1999, cadmium was detected at <1μg L^{-1} in the water supply in 5539 and at 2-10μg L^{-1} in 11 water samples. In all treated water, cadmium was detected at less than 1μg L^{-1}.

The estimated lethal oral dose of cadmium in humans is 350-3500 mg. With chronic oral exposure, the kidney appears to be the most sensitive organ. Cadmium affects the resorption function of the proximal tubules, the first symptom being an increase in the urinary excretion of low-molecular-weight proteins, known as tubular proteinuria. In Japan, chronic toxicity of cadmium or Itai-itai disease, an osteomalacia with various grades of osteoporosis accompanied by severe renal tubular disease, and low-molecular-weight proteinuria have been reported among people living in contaminated areas around the Zintsu river basin. Humans can be exposed to cadmium via food and drinking water from contaminated rivers polluted with the discharge of ore waste. A field survey found that the discharge of waste water in a mine contained 4 mg L^{-1}, and river water 10μg L^{-1} of cadmium. In river sediment, 0.16-0.18 ppm of cadmium was detected upstream of the mine, and 4.1-23.8 ppm downstream. The daily intake of cadmium in the most heavily contaminated areas amounted to 600-2000μg day^{-1}; in other less heavily contaminated areas, daily intakes of 100-390 μg day^{-1} have been found.

3.4. Chromium

Chromium is widely distributed in the earth's crust. It can exist in oxidation states of +2 to +6. Soil and rocks may contain small amounts of chromium, almost always in the trivalent state. Chromium and its salts are used in the leather tanning industry, the manufacture of catalysts, pigments and paints, fungicides, the ceramic and glass industries, and photography, and for chrome alloy and chromium metal production, chrome plating, and corrosion control.

The distribution of compounds containing chromium (III) and chromium (VI) depends on redox potential, pH, and the presence of oxidizing or reducing compounds. In the environment, chromium (VI) occurs mostly as CrO_4^{2-} or $HCrO_4^{-}$ and chromium (III) as $Cr(OH)_n^{(3-n)+}$. In water, chromium (III) is a positive ion that forms hydroxides and complexes, and is adsorbed at relatively high pH values. In surface water, the ratio of chromium (III) to chromium (VI) varies widely, and relatively high concentrations of the latter can be found locally. In general, chromium (VI) salts are more soluble than chromium (III) salts.

The average concentration of chromium in rainwater is in the range 0.2-1µg L^{-1}. Natural chromium concentrations of 0.04-0.5 µgL^{-1}L have been measured in seawater. In the North Sea, a concentration of 0.7 µg L^{-1} was found. Concentrations in surface waters are approximately 0.5-2µg L^{-1} and the dissolved chromium is content 0.02-0.3 µg L^{-1}. In general, the chromium concentration in groundwater is low (<1µg L^{-1}). In the Netherlands, a mean concentration of 0.7µg L^{-1} has been measured, with a maximum of 5 µg L^{-1}. In India, 50% of 1473 water samples from dug wells contained less than 2µg L^{-1}. In groundwater in the USA, levels of up to 50µg L^{-1} have been reported; in shallow groundwater, median levels of 2-10µg L^{-1} have been found. Most supplies in the USA contain less than 5µg L^{-1}.

Ingestion of 1-5 g of chromate results in severe acute effects such as gastrointestinal disorders, hemorrhagic diathesis, and convulsions. Death may occur following cardiovascular shock. In a long-term carcinogenicity study in rats given chromium (III) by the oral route, no increase in tumor incidence was observed. In rats, chromium (VI) acts as a carcinogen when inhaled, but there is limited evidence of carcinogenicity via the oral route. In epidemiological studies, an association has been found between exposure to chromium (VI) through inhalation and lung cancer. The IARC has classified chromium (VI) in Group 1 (human carcinogen).

3.5. Mercury

Mercury is used as a cathode in the electrolytic production of chlorine and caustic soda, in electrical appliances (lamps, arc rectifiers, mercury cells), in industrial and control instruments (switches, thermometers) in laboratory apparatus, in dental amalgams, and as a raw material for various mercury compounds. The solubility of mercury compounds in water varies; elemental mercury vapor is insoluble, mercury (II) chloride is readily soluble, mercury(I) chloride is much less soluble, and mercury sulfide has a very low solubility.

Contamination by mercury of water occurs from the discharge of mine waste and industrial waste. Pollution often yields inorganic and organic mercury contamination. Organic mercury in water occurs from the discharge of pesticides such as phenylmercury (II) acetate or effluent of acetaldehyde synthesis. However, methylation of inorganic mercury has been shown to occur in columns of fresh water and in seawater, and bacteria (*Pseudomonas* spp.) isolated from mucous material on the surface of fish and soil was able to methylate mercury under aerobic conditions. Some anaerobic bacteria that possess methane synthesize are also capable of methylating mercury. Environmental levels of methyl mercury depend on the balance between bacterial methylation and demethylation.

Levels of mercury in rainwater are in the range of 5-100 ng L^{-1}, but mean levels as low as 1 ng L^{-1} have been reported. Naturally occurring levels of mercury in groundwater and surface water are less than 0.5 µg L^{-1}, although local mineral deposits may produce higher levels in groundwater. An increase in the mercury concentration to $5.5\mu g\ L^{-1}$ was reported in wells on an island located near Tokyo, Japan, where volcanic activity is frequent.

Toxicity of mercury, the kidney is the main target organ for inorganic mercury, whereas methyl mercury mostly affects the central nervous system. In Japan, a degenerative neurological disorder caused by mercury poisoning, Minamata disease occurred in the 1950s to 1960s. Minamata disease, which is a typical example of the pollution-related health damage, was first discovered in 1956, around Minamata Bay in Kumamoto Prefecture, and in 1965, in the Agano River basin in Niigata Prefecture. Since the discovery of the disease, investigation of the cause has been made, and finally in 1968, the government announced its opinion that Minamata disease was caused by the consumption of fish and shellfish contaminated by methylmercury compound discharged from a chemical plant. It is known that Minamata disease is a disorder of the central nervous system and shows various signs and symptoms including sensory disturbance in the distal portions of four extremities, ataxia, concentric contraction of the visual field, etc. Industrial waste for acetaldehyde synthesis containing methyl mercury flowed into river and seawater, and contaminated aquatic organisms such as fish and shellfish through bioaccumulation. Damage to human health occurred directly through the consumption of contaminated fish and shellfish, and drinking water. A scheme of the production of acetaldehyde is shown in Figure 1.

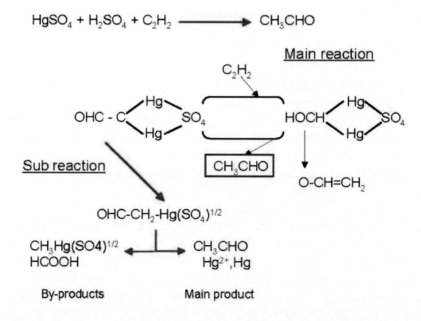

Figure 1: Methyl mercury production pathway by reaction of acetaldehyde synthesis

Acetaldehyde is synthesized in a reaction with acetylene, sulfuric acid and inorganic mercury as a catalyzer. During the reaction of acetylene and sulfuric acid, a low ratio of catalysis produces methyl mercury via a sub-reaction. Minamata disease is caused by river and sea water polluted with industrial waste containing inorganic mercury and methyl mercury. As the result of field survey, the total concentrations of mercury in fish and shellfish around the polluted area were as follows

---

Species	Hg (ppm) wet base
Gizzard Shad	1.62
White Croaker	8.4, 16.6
Sea Bass	16.6
Spanish Mackerel	8.72, 15.3
Gray Mullet	10.6
Crab	35.7, 23.9
Oyster	5.6
Short-necked Clam	20.0
Fishes(control area)	0.07-0.33
Shellfishes(control area)	0.08-0.1

Minamata Disease (1968)

Table 5: Concentration of total mercury in the fishes and shellfishes at Minamata bay

(Table 5); shellfish such as oysters and short-necked clams contained 5.6 to 20.0 ppm of mercury, crabs 14.0 to 35.7 ppm and fish such as sea bass 1.6 to 26.3 ppm. On the other hand, in aquatic organisms in unpolluted areas, total mercury levels ranged from 0.08 to 0.1 ppm in shellfish such as oysters and short-necked clams, and from 0.04 to 1.9 ppm in fish. It is clear that aquatic organisms living in mercury-polluted waters were contaminated by mercury after bioaccumulation.

Recently, mercury pollution in water has been reported around gold mining area such as the Amazon River basin in Brazil, Lake Victoria in Tanzania, Mindanao Island in the Philippines, Kalimantan Island in Indonesia, and Vietnam. The presence of mercury in water around gold mines is due to the use of mercury for producing Au-Hg amalgam. Gold particles such as gold dust collected from surface soil and river sediment in tropical rain forests produce an amalgam when metal mercury 2-7kg for every 1 kg of gold is added. Recovery the mercury in the amalgam by vaporization using a burner gives pure gold. The Vaporized mercury, however, pollutes the air and water. The results of a field survey found high concentrations of mercury in mud and sediments from polluted regions, and in fish living in polluted water.

3.6. Nickel

Nickel is used in a large number of alloys, including stainless steel, in batteries, chemicals, and catalysis, and in the electrolytic coating of items such as chromium-plated taps and fittings used for tap water. The nickel ion combines with oxygen-, nitrogen-, and sulfur-containing compounds and forms an extensive series of complexes in solution, including hydroxides, carbonates, carboxylic acids, phosphates, amines, and thiols. In aqueous solution, nickel occurs mostly as the green hexa-aquanicke l (II) ion, $Ni(H_2O)_6^{2+}$ containing ferrosulfide deposits. Nickel concentrations in drinking water around the world are normally below 20µg L^{-1}, although levels of up to several hundred micrograms per liter in groundwater and drinking water have been reported. Nickel concentrations in drinking water may be increased by natural or industrial nickel deposits or if leaching from nickel-chromium plated taps and fittings occurs. Levels of up to 1000μg L^{-1} have been reported in first-run water that had remained in the tap overnight.

Several epidemiological studies have suggested a risk of nasal, sinus, and lung from inhalation of nickel cancer in workers in the nickel-producing industry. The epidemiological data on respiratory exposure is sufficient to conclude that inhaled nickel sulfate is carcinogenic to humans. The IARC has classified nickel in group 1 (human damage).

3.7. Cyanide

Cyanides are used in many industrial processes such as electroplating, and the manufacture of coke and gas. A variety of chemical producers can be major sources of cyanide contamination in water. Cyanide sometimes occurs in surface water due to the discharge of plating waste, and causes the death of fish. Chlorination of potable water with a free chlorine residual under neutral or alkaline conditions will reduce the concentration of cyanide. Chlorination at more than pH 8.5 converts cyanides to innocuous cyanates, which can ultimately be decomposed to carbon dioxide and nitrogen gas. The cyanide ion is readily absorbed in animal species, and its highly poisonous effects are induced rapidly. A single dose of 50-60 mg is usually fatal in humans. Levels of 2.9-4.7 mg per day are regarded as non-injurious, owing to the highly efficient detoxification system in the human body whereby the cyanide ion is converted to the relatively non-toxic thiocyanate ion through the rhodanese and thiosulfate enzyme system. The WHO recommends a concentration of less than 0.07 mg L^{-1} in drinking water.

4. Inorganic Substances in Agricultural and Domestic Waste

Inorganic substances from effluent of agricultural and domestic waste often contaminate surface water or groundwater. Nitrogen and phosphate from the discharge of agricultural fertilizers and domestic waste can cause eutrophication. Nitric acid also from the discharge of agricultural fertilizers and from septic tanks for sewage can contaminate groundwater.

4.1. Nitrogen

Nitrate and nitrite are naturally occurring ions that are part of the nitrogen cycle. The nitrate ion (NO_3^-) is the stable form in oxygenated systems; although chemically unreactive, it can be reduced by microbial action. The nitrite ion (NO_2^-) contains nitrogen in a relatively unstable oxidation state; chemical and biological processes can further reduce nitrite to various compounds or oxidize it to nitrate.

Nitrate is used mainly in inorganic fertilizers. It is also used as an oxidizing agent and in the production of explosives and purified potassium nitrate for making glass.

In soil, fertilizers containing inorganic nitrogen and wastes containing organic nitrogen are first decomposed to give ammonia, which is then oxidized to nitrite and nitrate. In surface water, nitrification and denitrification may occur, depending on the temperature and pH.

Concentrations of up to 5 mg L^{-1} have been observed in rain water in industrial areas. The nitrate concentration in surface water is normally low (0-18 mg L^{-1}), but can reach high levels as a result of agricultural run-off, refuse dump run-off, or contamination with human or animal waste.

The natural nitrate concentration in groundwater under aerobic conditions is a few milligrams per liter and depends strongly on soil type and on geological conditions. As a result of agricultural activities, the nitrate concentration can easily reach several hundred milligrams per liter. For example, concentrations of up to 1500 mg L^{-1} were found in groundwater in an agricultural area of India. In Japan, when nitrogen fertilizer was used for growing vegetables at 500kg/h/year, nitrate as nitrogen (NO_3-N) was found in groundwater at a maximum concentration of 92mg L^{-1}. This concentration was 10 times the level in groundwater where no fertilizer was being used. The increasing use of artificial fertilizers, the disposal of wastes (particularly from livestock), and changes in land use are the main factors responsible for the progressive increase in nitrate levels in groundwater supplies over the last 20 years. In Denmark and the Netherlands, for example, nitrate concentrations are increasing by 0.2-0.3 mg L^{-1} per year in some areas. The NO_3-N concentration in groundwater has also increased.

Materials	Effluent(mg/L)	Treated W.(mg/L)	Remove R.(%)
BOD	53.1	0.6	99
Organic-N	6.84	0.38	94
NH$_4$-N	125	8.6	93
NO$_2$-N	0.20	0.89	-
NO$_3$-N	0.22	87.0	-
T-N	132	95.8	27
T-P	9.75	0.03	100

Table 6: Water qualities of effluent and treated water with penetrate treatment

Table 6 shows the change of NO_3-N in groundwater before and after the treatment of sewage. Sewage effluent contained high concentrations of organic nitrogen and NH_3-N, oxidized in soil as organic-N$-NH_3$-N$-NO_2$-N$-NO_3$-N. Groundwater after the treatment of sewage contained 87.6 mg L^{-1} of NO$_3$-N (438 times higher with sewage) while the concentration of NO_3-N in sewage was only 0.2 mg L^{-1}.

The toxicity of nitrate is thought to be a consequence of its reduction to nitrite. The major biological effect of nitrite in humans stems from its involvement in the oxidation of normal hemoglobin to methemoglobin, which is unable to transport oxygen to tissues. The result of reduced oxygen transport manifest clinically when methemoglobin concentrations reach 10% of hemoglobin concentrations; the condition, called methaemoglobinaemia, causes cyanosis and, at higher concentrations, asphyxia. Cases of methaemoglobinaemia related to low nitrate intake appear to be restricted to infants.

In studies in which a possible association between clinical cases of infant methaemoglobinaemia or subclinically increased methemoglobin levels and nitrate concentrations in drinking water was investigated, a significant relationship was usually found, with most clinical cases occurring at nitrate levels of 50 mg L^{-1} and above, and almost exclusively in infants under 3 months of age.

5. Inorganic Substances in the Water Supply

Inorganic substances in the water supply, purification treatment such as chlorination, and pipes leading from the purification plant to hydrants, all affect water quality aluminum act as coagulants in water treatment, and copper, lead, iron and zinc can contaminate tap water from pipes.

5.1. Aluminum

Aluminum is a widespread and abundant element, accounting for some 8% of the earth's crust. It is found as a normal constituent of soils and plants. Aluminum has many industrial and domestic applications. Its compounds are used as food additives, as antacids, and also in water treatment as flocculants. Aluminum may be present in water as a consequence of leaching from soil and rock. In a survey of aluminum in the USA, concentrations of 14-290 µg L^{-1} in groundwater and 16-1170µg L^{-1} in surface water were reported. In the UK, concentrations of 200-300µg L^{-1} were associated with low pH levels and 400-600µg/L with afforested catchments. In water in the Tokyo metropolitan area of Japan, values ranged 10-1200µg L^{-1} in surface water and 20-230µg L^{-1} in lake and reservoir water. Aluminum salts are used as coagulants in water treatment. The residual aluminum concentration is a function of the level in the source water, the amount of aluminum coagulant used, and the efficiency of filtration of the aluminum floe. Where residual concentrations are high, aluminum may be deposited in the distribution system, and a gradual reduction with distance from the treatment plant may then be observed.

Aluminum in water is suspected to contribute to Alzheimer disease, the first recognizable symptoms of which are memory lapses, disorientation, confusion, and frequent depression. Aluminum is one of numerous causal factors that have been proposed for this disease. The brains of Alzheimer patients appear to have elevated levels of aluminum in regions containing large numbers of neurofibrillary tangles. Few attempts have been made to study the relationship between Alzheimer disease and the concentration of aluminum in drinking water. For example, an epidemiological study has been carried out in the UK in which the exposure to aluminum from drinking water was calculated from data provided by local water undertakings. Districts in which aluminum concentrations in drinking water exceeded 0.01 mg L^{-1} (four subsets; 0.02-0.04 mg L^{-1}, 0.05-0.07 mg L^{-1}, 0.08-0.11 mg L^{-1}, and >0.11 mg L^{-1}) were found to have an approximately 50% greater incidence of Alzheimer disease as compared with those in which aluminum concentrations were below 0.01 mg L^{-1}. Epidemiological and physiological evidence does not at present support a causal role for aluminum in Alzheimer disease. Therefore, the WHO guideline of 0.2 mg L^{-1} provides a compromise between the practical use of aluminum salts in water treatment and the discoloration of distributed water.

5.2. Copper

Copper is an important heat and electrical conductor. It is also used for water pipes, roof coverings, household goods, and chemical equipment, and in many alloys. Copper oxides, chlorides, sulfates, ehtanoates, bromides, and carbonates are widely used in pest control as inorganic dyes, as feed additives, in seed disinfectants, and in electroforming. Monovalent copper is unstable in aqueous solution. Copper (II) forms complexes with both inorganic and organic ligands such as ammonium and chloride ions and humic acids.

Natural copper concentrations in drinking water are a few micrograms per liter. Depending on hardness, pH, anion concentrations, oxygen concentrations, temperature, and the pipe system, water from copper pipes may contain several milligrams of copper per liter. Concentrations of copper in raw and tap water in Japan (1999) are summarized in Figure 2.

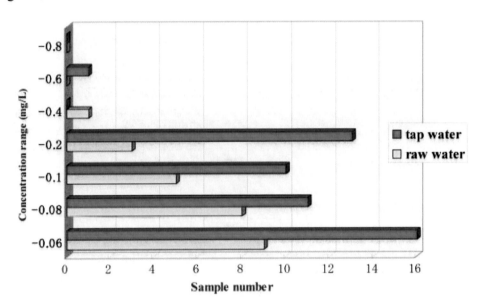

Figure. 2: Concentration of copper in raw and tap water

Some 66 of 5551 water samples (1.2%) contained more than 0.04 mg L^{-1} of copper. However, 125 of 5705 tap water samples (2.2%) contained a concentration of more than 0.04 mg L^{-1}. Copper is an essential element for humans and animals. Acute gastric irritation may be observed in some individuals at concentrations in drinking water above 3 mg L^{-1}, and long-term ingestion can give rise to liver cirrhosis. Copper can cause taste problems and furnish water blue.

5.3. Lead

Lead is the commonest of the heavy elements, accounting for 13 mg kg^{-1} of the earth's crust. Lead is used in the production of lead acid batteries, solder, alloys, cable sheathing, pigments and plastic stabilizers. The almost universal use of lead compounds in plumbing fittings and as solder in water distribution systems is important. Lead pipes may be used in older distribution systems and plumbing. Lead is present in tap water to some extent as a result of its dissolution from natural sources but primarily from household plumbing systems in which the pipes, solder, fittings, or service connections to homes contain lead. The amount of lead dissolved from the plumbing system depends on several factors, including the presence of chloride and dissolved oxygen, pH, temperature, water softness, and standing time of the water, soft, acidic water being the most plumbosolvent. The level of lead in drinking water may be reduced by corrosion control measures such as the addition of lime and the adjustment of the pH in the distribution system from <7 to 8-9.

A recent review of lead levels in drinking water in the USA found the geometric mean to be 2.8 μg L^{-1}. A recent study in Canada (Ontario) found that the average concentration of lead in water actually consumed over a one week sampling period was in the range 1.1-30.7μg L^{-1}, with a median level of 4.8μg L^{-1}. Lead levels in drinking water from a survey in Japan in 1999 are summarized in Figure 3.

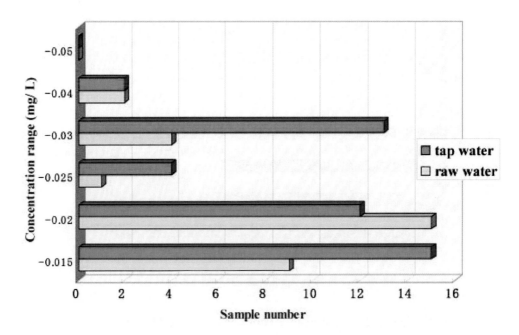

Figure 3: Concentration of lead in raw and tap water

Of 5551 water samples, 69 (1.6%) contained more than 0.01 mg L^{-1} of lead. However, 126 of 5705 tap water samples (2.2%) had more than 0.01 mg L^{-1}. The concentration of lead in raw and tap water ranged from 0.02 to 0.04 mg L^{-1}. At the concentration of 0.03mg L^{-1}, tap water had about 3 times more lead than raw water. It is clear from the concentrations of lead in raw and tap water (Figure.2), that the lead in tap water is mainly dissolved from lead pipes.

6. Radioactive Material

Natural ionizing radiation, to which everyone is exposed, includes cosmic rays and products of the decay of radioactive elements in the earth's crust and atmosphere. Part of the terrestrial radiation dose is from sources external to the body, and part is due to the inhalation and ingestion of radioactive elements in air, food and water. Minute traces of radioactivity are normally found in all drinking water. The concentration and composition of these radioactive constituents vary from place to place, depending principally on the radiochemical composition of the soil and rock strata through which the raw water may have passed. Many natural and artificial radio nuclides have been found in water, but most of the radioactivity is due to a relatively small number of nuclides and their decay products. By far the largest contribution to the radioactivity in drinking water comes from potassium-40, which is present as a constant percentage of total potassium. Only a small percentage of the total potassium-40 body burden, however, comes from drinking water.

Water contamination by radioactive materials has occurred with accidents at atomic power plants.

On April 26, 1986, a major accident, determined to have been a reactivity (power increase) accident, occurred at Unit 4 of the nuclear power station at Chernobyl, Ukraine, in the former USSR. The accident destroyed the reactor and released massive amounts of radioactivity into the environment. A total of about 12×10^{18} Bq of radioactivity was released.. Main nuclear decay products were iodide 131 and cesium 137. In European countries, radial rays levels increased about 100-fold during late April to early May and contaminated drinking water and foods such as milk, meat and vegetable. Radioactive radiation also affected Asia. Variations of gross beta activity in rain water in Hokkaido and Tokyo in Japan are shown in Figure 4.

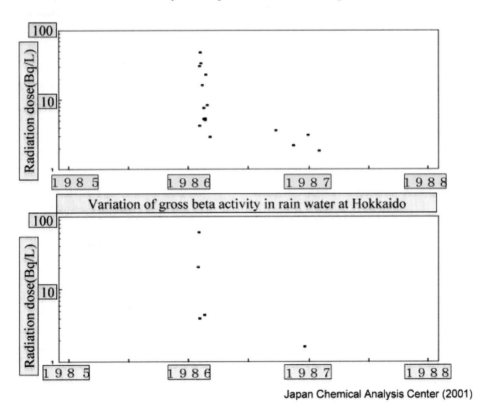

Variation of gross beta activity in rain water at Hokkaido

Japan Chemical Analysis Center (2001)

Figure 4: Variation of gross beta activity in rain water at Tokyo, Japan

The rain water at Hokkaido located in the northern part of Japan had gross beta activity readings of 2.9 to 47.9Bq L^{-1} 10 days after the Chernobyl accident. Rain water in Tokyo detected showed gross beta activity values of 4.0 to 60.9Bq/L after 7 days. In rain water collected in Tokyo, the iodide 131 concentration was 1700 pCi L^{-1} and the gross gamma activity concentration was about 8000pCi L^{-1} 6 and 7 days after the Chernobyl accident. Exposure to radiation in drinking water can have teratogenic, genetic, somatic and carcinogenic effects.

Glossary

Alzheimer disease:	Alzheimer Disease (AD) is a brain disorder that seriously affects a person's ability to carry out daily activities. The onset of AD is typically gradual, and the first signs of it may be attributed to old age or ordinary forgetfulness.
Chernobyl:	Chernobyl nuclear power plant in the Ukraine (former Soviet Union). The power plant accident appeared on April 26, 1986. The resulting steam explosion and fire released at least five percent of the radioactive reactor core into the atmosphere and downwind.
Bq/L:	Becquerel(Bq) is unit of radioactivity derived SI unit. 1 Bq represents one disintegration or other nuclear transformation per second. Bq has equivalent 2.703 x 10^{-11} curies
Itai-itai disease:	Health effects as a result of cadmium pollution of water environment. Itai-itai disease which disorder are osteomalacia with various grades of osteoporosis accompanied by severe renal tubular disease.
Methaemoglobinaemia:	Biological effect such as cyanosis and asphyxia observed. Condition as 10% or above concentrations of methaemoglobin in haemoglobin. Nitrite in humans oxidize normal hemoglobin to metohaemoglobin
Methylation:	Condition to change in the organic methylated compound from the inorganic compound (i.e. Inorganic mercury to methyl mercury)
Methyl mercury:	Methylated mercury. Some bacteria isolated from mucous material are able to methylate mercury under aerobic conditions. This compound also produce by acetaldehyde synthesis as by-product.
Minamata disease:	Minamata disease is a disorder of the central nervous system and shows various signs and symptoms including sensory disturbance in the distal portions of four extremities, ataxia, concentric contraction of the visual field. Minamata disease was caused by the consumption of fish and shellfish contaminated by methylmercury compound discharged from a chemical plant.

Bibliography

Akagi H., Kinjo Y., Branches F., Malm O., Harada M., Pfeiffer W.C. and Kato H.(1995)Methyl mercury pollution in the Amazon, Brazil. The Sci. of Total Environ. 175, 85-95[This article shows water pollution with methyl mercury in Amazon river region]

ATSDR (1993) Toxicological profile for fluorides, hydrogen fluoride, and fluorine. Atlanta, Georgia, US Department of Health and Human Services, Agency for Toxic Substances and Disease Registry (TP-91/17). [This article shows toxicity of fluoride including concentration of fluorides in surface water]

Butterwick L, De Oude N, & Raymond K (1989) Safety assessment of boron in aquatic and terrestrial environments. Ecotoxicol Environ Saf, 17, 339-371. [This article contains boron concentration of various surface water]

Camargo JA (1996a) Comparing levels of pollutants in regulated rivers with safe concentrations of pollutants for fishes: a case study. Chemosphere, 33(1): 81–90. [This article shows a result of fluorides monitoring in the Duraton River, Spain]

Elwood P.C., Abernethy M., Morton M.(1974) Mortality in adults and trace elements in water, Lancet 2, 1470-1472 [This article shows associations between the barium content of drinking water and mortality from cardiovascular disease]

Hirayama N, Maruo M, Obata H, Shiota A, Enya T, & Kuwamoto T (1996) Measurement of fluoride ion in the river-water flowing into Lake Biwa. Water Res, 30(4): 865–868. [This article reported fluoride concentration in the river water flowing into Lake Biwa, Japan]

Jacks G, Sharma VP (1983) Nitrogen circulation and nitrate in groundwater in an agricultural catchment in southern India, Environmental geology, 5, 61-64 [This article shows contamination of groundwater by agricultural fertilizer.]

Japan Chemical Analysis Center (2001), Radioactivity Survey Data in Japan,

http://www.kankyo-hoshano.go.jp/en/index.html [This data base shows gross beta activity in atmospheres of 47 sites in Japan]

Magara Y et al.(1989) Effects of volcanic activity on heavy metal concentration in deep well water.Technical papers Water Nagoya'89; 7th Regional conference and exhibition of Asia-Pacific. International Water Supply Association, 411-419 [This paper reported that Izu Oshima Island located Tokyo metropolitan area, deep well water are contained inorganic mercury effected by volcanic activity]

Minamata Disease (1968) Study group of Minamata Disease, Kutsuna M.,ed., Kumamoto University, Japan [This report shows review of field survey, clinical and epidemiological studies on chronic Minamata disease]

Minamata Disease, Key Topics, Ministry of the Environment, Japan, http://www.env.go.jp/en/index.html/ [This topics shows history and measure of Minamata Disease]

Nair KR, Manji F, & Gitonga JN (1984) The occurrence and distribution of fluoride in groundwaters in Kenya. East Afr J Med, 61: 503–512 [This article reported high concentration of fluoride in many of lakes at Kenya]

Robert N,(1974) Some facts about cadmium. AMBIO, 3, 56-66 [This article shows reviewing the toxicity of cadmium and health effects including Itai-itai disease]

Schroeder H.A. and Kramer L.A.(1974) Cardiovascular mortality, municipal water, and corrosion, Arch. Of environ. Health, 28, 303-311 [This article shows associations between the barium content of drinking water and mortality from cardiovascular disease]

Skjelkvåle BL (1994) Water chemistry in areas with high deposition of fluoride. Sci Total Environ, 152: 105–112. [This article shows a high concentration of fluoride detection in the stream of close to aluminum smelters]

Walton G.(1951) Survey of literature relating to infant methemoglobinemia due to nitrate contaminated water, American J. Public Health, 41, 986-996 [This article described about infant methemoglobinemia caused nitrate in drinking water]

Weast RC, Astle MJ, & Beyer WH ed. (1985) CRC handbook of chemistry and physics, 69th ed. Boca Raton, Florida, CRC Press, Inc., pp B/77, B/129. [This book contains average concentration of boron in ocean water]

WHO, IPCS(The international programme on chemical safety), Environmental health criteria(EHC) No.194 Aluminum (1997) [This book describes about environmental health criteria for aluminum including physical and chemical properties, human and environmental exposure, metabolism, effects on animals and human, evaluation of human health risks and effects on environment]

WHO, IPCS (The international programme on chemical safety), Environmental health criteria(EHC) No.224 Arsenic (2nd edition) (2001) [This book describes about environmental health criteria for arsenic including physical and chemical properties, human and environmental exposure, metabolism, effects on animals and human, evaluation of human health risks and effects on environment]

WHO, IPCS (The international programme on chemical safety), Environmental health criteria(EHC) No.204 Boron (1998) [This book describes about environmental health criteria for boron including physical and chemical properties, human and environmental exposure, metabolism, effects on animals and human, evaluation of human health risks and effects on environment]

WHO, IPCS (The international programme on chemical safety), Environmental health criteria(EHC) No.61 Chromium (1988) [This book describes about environmental health criteria for chromium including physical and chemical properties, human and environmental exposure, metabolism, effects on animals and human, evaluation of human health risks and effects on environment]

WHO, IPCS (The international programme on chemical safety), Environmental health criteria(EHC) No.200 Copper (1998) [This book describes about environmental health criteria for copper including physical and chemical properties, human and environmental exposure, metabolism, effects on animals and human, evaluation of human health risks and effects on environment]

WHO, IPCS (The international programme on chemical safety), Environmental health criteria(EHC) No.227 Fluorides (2002) [This book describes about environmental health criteria for fluorides including physical and chemical properties, human and environmental exposure, metabolism, effects on animals and human, evaluation of human health risks and effects on environment]

WHO, IPCS (The international programme on chemical safety), Environmental health criteria(EHC) No.85 Lead environmental aspect (1989) [This book describes about environmental health criteria for lead including physical and chemical properties, human and environmental exposure, metabolism, effects on animals and human, evaluation of human health risks and effects on environment]

WHO, IPCS (The international programme on chemical safety), Environmental health criteria(EHC) No.118 Mercury inorganic (1991) [This book describes about environmental health criteria for inorganic mercury including physical and chemical properties, human and environmental exposure, metabolism, effects on animals and human, evaluation of human health risks and effects on environment]

WHO, IPCS (The international programme on chemical safety), Environmental health criteria(EHC) No.101 Methyl mercury (1990) [This book describes about environmental health criteria for methyl mercury including physical and chemical properties, human and environmental exposure, metabolism, effects on animals and human, evaluation of human health risks and effects on environment]

WHO, IPCS (The international programme on chemical safety), Environmental health criteria(EHC) No.108 Nickel (1991) [This book describes about environmental health criteria for nickel including physical and chemical properties, human and environmental exposure, metabolism, effects on animals and human, evaluation of human health risks and effects on environment]

WHO, Guidelines for drinking-water quality, second edition volume 2 (1996)

Yuko Nakamura and Takashi Tokunaga(1996) Antimony in the aquatic environment in north Kyushu district of Japan, Water Science and Technology Vol 34 No 7-8 pp 133–136 [This article reported that antimony concentrations in the aquatic environment in north Kyushu district of Japan found in the range 0.0 to 0.8 μ g/l]

Zdanowicz JA et. al. (1987) Inhibitory effect of barium on human caries prevalence, Community dentistry and oral epidemiology, 15, 6-9 [This article reported that epidemiological study about prevalence of dental caries with barium in drinking water]

Biographical Sketch

Yoshiteru Tsuchiya is a Lecturer of Faculty of Engineering Kogakuin University, where he has been at present post since 2000. He obtained the Bachelor Degree in Meiji Pharmaceutical College in 1964. He worked for Department of environmental health in Tokyo Metropolitan Research Laboratory of Public

Health until 2000. In the meantime, he obtained the Ph.D in Pharmaceutical Sciences from Tokyo University. From 2002 to 2003, he worked for the Yokohama National University Cooperative Research and Development Center as Visiting Professor.

Dr. Tsuchiya has written and edited books on risk assessment and management of waters. He has been the author or co-author of approximately 70 research articles.

He is the member of Japan Society on Water Environment, Pharmaceutical Society of Japan, Japan Society for environmental Chemistry, Japan Society of Endocrine Disrupters Research, and International Water Association(IWA).

MICROBIAL/BIOLOGICAL CONTAMINATION OF WATER

Yuhei Inamori
Executive researcher, National Institute for Environmental Studies, Tsukuba, Japan

Naoshi Fujimoto
Assistant Professor, Faculty of Applied Bioscience, Tokyo University of Agriculture, Tokyo, Japan

Keywords: hazardous microorganism, drinking water, biological treatment, infection

Contents

Summary

Microorganisms that affect human health greatly include pathogenic bacteria, pathogenic viruses, pathogenic protozoa, and cyanobacteria.

Japanese regulations concerning bacteria levels for drinking water safety require that in 1 ml of test water, the number of colony of general bacteria created is under 100, that no total coliforms are detected, and that residual concentration of free chlorine at the faucet is 0.1 mg $\cdot$ l^{-1} or more.

There are approximately 100 kinds of viruses known to infect humans, and many viruses originating from human excrement are contained in sewage water, such as polio, coxsackie, adeno, and the influential hepatitis virus.

Growth of cyanobacteria is caused when nitrogen and phosphorus concentration increase by contamination with domestic sewage, industrial sewage, and livestock sewage. Thus it is particularly effective to prevent the production of cyanobacterial toxin by reducing nitrogen and phosphorus in the wastewater.

Protozoa and metazoa contribute removal of pathogenic bacteria, pathogenic virus, cyanobacteria and pathogenic protozoa by predation in biological treatment.

1. Introduction

For human beings, the critical issue when using water is hygiene. More than 4 million people die of illnesses contacted through microorganisms, and most cases are caused by water contaminated by microorganisms. There are many forms of water use in daily life, but the greatest threat to human life occurs when there is direct contact between water and human beings, for example bathing spots where sewage is mixed into the water, office buildings that treat and recycle waste water from toilets for reuse, and water works that use river water as the water supply source. In such cases, microorganisms that affect human health greatly include pathogenic bacteria, pathogenic viruses, pathogenic protozoa, and cyanobacteria (Table 1).

	Microorganisms	**Disease**
Bacteria	*Salmonella typhi*	typhoid
	Salmonella choleraesuis	typhoid, gastroenteritis
	Salmonella enteritidis	typhoid, gastroenteritis
	Shigella sp.	dysentery
	Vibrio cholerae	cholera
	Camplobacter jejuni	enteritis
	intestinal pathogenic coliform	gastroenteritis
	Mycobacterium tuberculosis	tuberculosis
Virus	*rotavirus*	gastroenteritis
	poliovirus	infantile paralysis
Protozoa	*Cryptosporidium*	typhoid
	Giardia	typhoid
	Entamoeba	dysentery
Algae	*Microcystis*	liver disorder
	Aphanizomenon	nervous disorder
	Anabaena	nervous disorder
	Cylindrospermopsis	liver disorder

Table 1: Harmful microorganism in water environment

These pathogenic microorganisms reproduce within the body and infect the body. Cyanobacteria that produce toxic substances, on the other hand, do not reproduce inside the body, but infect the body when more than the tolerable volume of toxic substances that it produces is ingested through contaminated tap water. Water contamination caused by pathogenic microorganisms and microorganisms that produce toxic substances has become a serious problem. Toxic microorganisms are becoming increasingly common in eutrophic or polluted water, and such problems must be solved. This part explains separately at bacteria, viruses, and pathogenic protozoa that cause contamination, and also discusses the elimination of pathogenic microorganisms during biological treatment of sewage water.

2. Bacteria

Bacteria that are currently targeted by Japanese water quality standards are general bacteria, total coliforms, and fecal coliforms. The standards are determined depending on how the water is used. One of the most widely used of the various water standards in Japan is the number of total coliforms. The number of general bacteria is regulated by the Waterworks Law water quality standard (Table 2), while fecal coliforms are regulated only for public recreational waters. This is because in recreational waters, there is a high possibility of water being ingested orally, and so it is important to prevent contamination by pathogenic microorganisms to maintain water quality. It should also be noted that when treated water from sewage works is recycled for use as water for sprinklers or landscaping, where human beings may come in direct contact with the water, the water quality standard requires that no total coliforms be detected.

Standard	Objective bacteria	value
World Health Organization, guideline for drinking water quality	Total coliform or	0・100ml-1
	fecal coliform	
Drinking water quality based on waterworks law in Japan	General bacteria	Less than 100・l-1
	Total coliform	No detection
Environmental standards for lakes and reservoirs in Japan	Total coliform	Less than 1000 MPN・100ml-1 for drinking water resource and bathing
Effluent standard based on water pollution control law in Japan	**Total coliform**	3000・ml-1

Table 2: Standard for bacteria in drinking water, water environment and effluent

As seen from these examples, water quality standards depend on how the water is to be used, but in general when human beings come in direct contact, the water quality is confirmed using fecal coliform numbers. When lakes, reservoirs and ground water are contaminated by pathogenic bacteria, or are not completely sterilized by the water purification treatment, they may become a source of infectious diseases. Typical water-borne infectious diseases include cholera and dysentery, which took the lives of many from the 19th century to the first half of the 20th century. Modern water systems and water sterilization efforts have slowed the spread of such water-borne infectious diseases, but there was a major cholera outbreak in 1991 which led to many deaths in Central and South America, and in Africa. The cause is believed to have been poor treatment systems of waterworks and sewage sewer, and deteriorating living conditions. *Vibrio cholerae* is the bacteria causing cholera, while *Shigella* causes dysentery. Pathogenic colon bacillus, *Campylobacter*, *Clostridium*, *Salmonella*, and *Staphylococcus* are some of the bacteria known to cause water-borne infectious diseases. All of these pathogenic bacteria infect the human intestine, are released into the outside environment with excrement, pass through the treatment process, and are released into rivers, lakes and reservoirs. *Escherichia coli* O157 and other fecal coliforms are increasingly causing water-borne infectious diseases in recent years in Japan. O157 is part of the pathogenic colon bacillus which produces verotoxin and causes hemorrhagic colitis, an intense form of diarrhea, hemolytic uraemia syndrome, and thrombotic thrombocytopenic purpura, in cases leading to death. The prevention of infection by such pathogenic bacteria is of utmost importance. Japanese regulations concerning bacteria levels for drinking water safety are detailed in ministry orders on water quality standards, and require that in 1 ml of test water, the number of colony of general bacteria created is under 100, that no total coliforms are detected, and that residual concentration of free chlorine at the faucet is 0.1 mg $\cdot$ l^{-1} or more.

3. Viruses

Viruses differ from bacteria and other microorganisms in that they hold only DNA or RNA as its nucleic acid (other microorganism cells contain both DNA and RNA), have only protein to protect the nucleic acid, and increase only in living cells. They measure from 10 to 450 nm. Intestinal viruses can cause symptoms seen in summer colds, fever, and diarrhea. There are approximately 100 kinds of viruses known to infect humans, and many viruses originating from human excrement are contained in sewage water, such as polio, coxsackie, adeno, and the influential hepatitis virus. Therefore, it is possible that these viruses also exist in rivers, lakes, and other natural water bodies where treated sewage water is discharged. Intestinal viruses that affect humans normally do not proliferate in natural waters, but they are not easily decomposed in the natural environment, and remain for a long period in the waters, so it is best to perform inactivation in the sewage treatment process. Moreover, viruses that have been adsorbed by sludge produced in the biological wastewater treatment are not completely inactivated, and can contaminate ground water when sludge is disposed of in wooded or agricultural areas, so it is dangerous to dispose of raw sludge into the environment. Chemicals used to inactivate viruses include chlorine and ozone. Ozone has no residual-acting characteristics, so it would be effective to use ozone, which has a strong virucidal effect, to efficiently eliminate viruses during water treatment, and then use low-concentration chlorine to finish the process.

4. Pathogenic Protozoa

The most common pathogenic protozoa that cause contamination are *Cryptosporidium* and *Giardia*. *Cryptosporidium* is a parasitic protozoa belonging to sporozoea coccidiida. Of the two types that infect animals, the larger *Cryptosporidium muris* lives in the stomach, and its oocyst measures 6.6-7.9×5.3-6.5μm, while the smaller *C. parvum* lives in the intestine and its oocyst measures 4.5-5.4×4.2-5.0μm. Both are elliptical in shape, and *C. parvum* affects humans, though it is reported that immunodeficient patients can also be infected by *C. muris*. It is known that the immune system normally increases serum complement activities within the body to naturally cure this infection, but this function is impaired in immunodeficient patients. *Cryptosporidium* infections have been reported several times in recent years, including a 1993 outbreak in Milwaukee, Wisconsin infecting 1.6 million people, of which 403,000 people showed symptoms, and more than 400 died. *Cryptosporidium* exists as oocysts in environment, and cannot grow. Once the oocyst enters the body of a mammal through the mouth, it creates oocysts through non-sexual and sexual reproduction in the alimentary canal. The grown oocysts exit the body through excrement, and become an infection source for other animals, while some exert sporozoids inside the alimentary canal, causing self-infection. The infection is caused by the oocyst which is discharged from the host, for example cattle, horses, pigs, dogs, cats, and mice; the communication means is oral ingestion; the incubation period is approximately 4-10 days; symptoms include abdominal pain and water-like diarrhea which lasts around 3-7 days; and the oocyst maintains its infectiousness for a long period by not becoming inactive for 2-6 months in humid air after it is discharged into the environment. Infectiousness is lost after 30 minutes at $-20^{\circ}C$, or 1-4 days at room temperature or in dry conditions.

In the US, the number of *Cryptosporidium* existing in environmental waters not affected by man-made contamination, and which has contact with wild animals, is 0.04 oocysts · l^{-1} in pond water, 0.003 oocysts · l^{-1} in ground water, 10^2-10^4 oocysts · l^{-1} in untreated sewage water, 10^1-10^3 oocysts · l^{-1} in water disinfected with chlorine after the activated sludge process (removal rate 96.6%), which is reduced to a tenth (removal rate 99.9%) through sand filtration. To inactivate 90% of the oocysts, 80 mg · l^{-1} of free chlorine for 90 minutes, or 1 mg · l^{-1} of ozone for 5 minutes is required, meaning that the disinfecting treatment commonly conducted in water treatment facilities using free chlorine is not sufficient to inactivate *Cryptosporidium*. The USEPA sets tolerable concentration levels of toxic organisms which are calculated to prevent infection of more than one in every 10,000 a year, and its *Cryptosporidium* concentration rate in drinking water that would infect one person in 10,000 in one year is 0.000033 oocysts · l^{-1}. The conventional rapid sand filtration method eliminates approximately 99.9% of the *Cryptosporidium*, so that if more than 3.3 oocysts of *Cryptosporidium* are contained in 100 liters of untreated water, it would not meet the tolerable concentration. It is extremely important therefore to appropriately operate the coagulation-sedimentation treatment to prevent *Cryptosporidium* infection, and thoroughly conduct water treatment by incorporating membrane separation or other methods. It is important to use highly sophisticated sewage treatment facilities for wastewater to be discharged to the drinking water resource, or to take appropriate measures at the drinking water resource, such as changing the intake of drinking water.

5. Cyanobacteria

Anabaena flos-aquae	anatoxin (neurotoxin)
	microcystin (hepatotoxin)
Aphanizomenon flos-aquae	aphantoxin (neurotoxin)
Cylindrospermopsis raciborskii	cylindrospermopsin (hepatotoxin)
Microcystis aeruginosa	microcystin (hepatotoxin)
Nodularia spumigena	nodularin (hepatotoxin)
Oscillatoria agardhii	microcystin (hepatotoxin)

Table 3: Toxic cyanobacteria and its toxin

It has been confirmed that toxic cyanobacteria species exist in the water bloom caused by eutrophication of lakes and reservoirs. Some common cyanobacteria that produce toxins are *Anabaena*, *Aphanizomenon*, *Cylindrospermopsis*, *Microcystis*, *Nodularia*, *Nostoc*, and *Oscillatoria*. A hepatotoxin microcystin is frequently detected in eutrophicated lakes and reservoirs worldwide. Table 3 shows common types of toxic cyanobacteria and its toxic substances.

Toxic cyanobacteria are causing the death of livestock and humans. The tap water of one Brazilian hospital was contaminated by microcystin, leading to the death of more than 50 people. A microcystin is a cyclic peptide made of 7 amino acids (Figure 1).

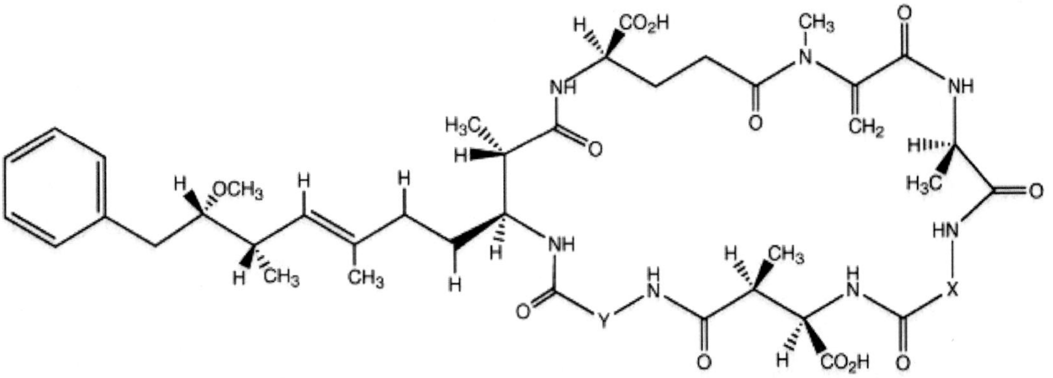

	X	Y
RR	arginine	arginine
YR	tyrosine	arginine
LR	leusine	arginine

Figure 1: Structure of microcystin

Different positions of amino acids X and Y make three different type of microcystin: microcystin-RR, microcystin-YR, and microcystin-LR. There are close to 60 microcystin congeners. Microcystin-LR is believed to be the most toxic, and the LD_{50} for mice is reported to be $50 \mu g \cdot kg^{-1}$. There are two types of *Microcystis*: toxic containing microcystin, and non-toxic without microcystin. The amount of microcystin in lakes and reservoirs is known to differ in the same water body, depending on the place, depth and time. Another characteristic is that *Microcystis* and other cyanobacteria have strong buoyancy and accumulate at the shore easily when blown together by the wind, so that the microcystin concentration differs greatly in different parts of the water body. There are many reports on microcystin amounts in various countries, for example 0.2-0.4 mg $\cdot$ g dry weight^{-1} in Japanese lakes and reservoirs, 0.7-0.8 mg $\cdot$ g dry weight^{-1} in reservoirs in Thailand, and 0.05-0.415 mg $\cdot$ g dry weight^{-1} in reservoirs in South Africa. Figure 2 shows the existing microcystin amount in Japanese lakes and reservoirs, where it is produced mostly in summer, with a greater amount of microcystin-RR and microcystin-LR.

An effective method of eliminating microcystin from drinking water is oxidation treatment with ozone, adsorption treatment with activated carbon, and biological treatment. In the biological treatment, it is reported that protozoa and metazoa predate the cyanobacteria, and the microcystin that elutes out of cell is ingested and decomposed by bacteria (Figure 3). Microcystin normally exists inside the cell, but dissolves into the water when it dies, so manager of drinking water treatment should pay close attention to microcystin levels at each stage of the treatment process during the time of year when cyanobacteria die. Growth of cyanobacteria is caused when nitrogen and phosphorus concentration increase by contamination with domestic sewage, industrial sewage, and livestock sewage. Thus it is particularly effective to prevent the production of cyanobacterial toxin by reducing nitrogen and phosphorus levels in the wastewater.

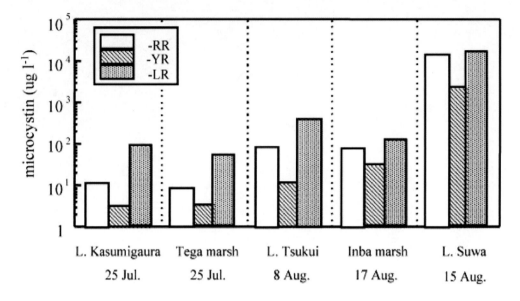

Figure 2: Microcystin concentration in Japanese lakes and marshes in summer

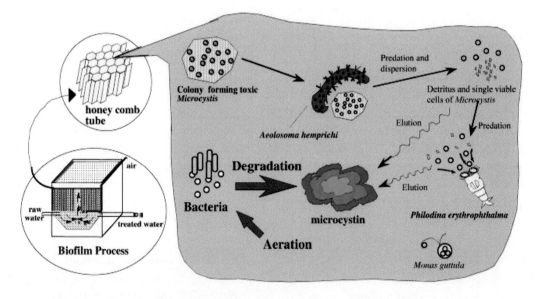

Figure 3: Diagram of biodegradation process for microcystin by protozoa, metazoa and bacteria

6. Dinoflagellates

Some dinoflagellates produce toxic substances, and fish and shellfish can be damaged or accumulate toxins within themselves, which can cause intoxication symptoms in humans when eaten. Dinoflagellates produce ciguatoxin, saxitoxin, okadaic acid, and other toxins. One dinoflagellate which has become a major problem recently is *Pfiesteria piscicida*, one of the approximately 2000 algae causing red tide which exist worldwide. *P. piscicida* is believed to have caused the death of many fish in the mouth of river in North Carolina and Maryland. US law requires a level 3 laboratory to deal with this microorganism. The toxin produced by *P. piscicida* destroys fish skin, and damages the neural system and vital organs. It also produces a volatile toxin, which causes health problems in humans, can turn immediately from being non-toxic to toxic in a short period when they find fish. It is also reported that fish eggs do not hatch and young mussels lose the ability to close their shells when *P. piscicida* exist in the water. *P. piscicida* can be moved from one country to another through the ballast water of ships, and close attention is required when large amounts of ballast water is being moved at ports and loading docks. It can also be carried with raw seafood, such as imported sea urchins, and it is important to confirm whether there have been any reports of *P. piscicida* where it came from. *P. piscicida* increases drastically by consuming the large quantity of algae that is grown with eutrophication. Thus it is important to control the nitrogen and phosphorus discharged to the public water bodies to prevent the increase of toxic dinoflagellate.

7. Removal of pathogenic microorganisms by biological treatment

Microorganisms interact with each other in biological treatment through impartial relationships, competition, symbiosis, synergism, commensalism, amensalism, parasitism, and prey-predator relationships. Processes used in sewage treatment, such as the standard activated sludge process, anaerobic-aerobic activated sludge process (AO, A$_2$O process), activated sludge process with flocculants, biofilm process, and immobilization process, all use microorganismic ecosystems which is composed of food webs of bacteria, fungus, protozoa and microscopic metazoa such as rotatoria, oligochaeta. An imbalance in this ecosystem may cause treatment failure, including bulking, dispersion of sludge, and milkiness in the treated water. The prey and predator system is extremely important because it has roles such as predation of filamentous microorganisms and control of bulking, predation of dispersed microorganisms such as *Cryptosporidium* and O-157 pathogenic bacteria, keeping high transparency of treated water, and stabilization of the treatment efficiency. The predator organisms, which play an important role in the biological treatment reactor, include *Vorticella* (ciliata), *Carchesium, Cyclidium, Trithigmostoma, Amoeba* (sarcodina), *Philodina* (rotatoria), *Rotaria, Aeolosoma* (oligochaeta), and others. These protozoa and metazoa prey ranging in size from dispersed *Escherichia coli*, approximately 1μm, to cohered microorganisms, approximately 100μm, and secrete biopolymers to cohere micro-suspensions. Thus it helps greatly in improving solid-liquid separability, clearing of treated water, and reduction of sludge. When interaction among microorganisms in the activated sludge process is performing well, and the prey-predator relationship is balanced, the transparency of treated water can reach 2 meters, and almost no total coliforms will be detected even without chlorine disinfection (Table 4). It is therefore important to maintain the protozoa and microscopic metazoa density in the biological treatment reactor.

In an experiment to confirm predatory characteristics using *Escherichia coli*, all predators used – *Philodina erythrophthalma, Cyclidium glaucoma, Tetrahymena pryformis, Colpidium campylum*, and *Aeolosoma hemprchi* – were effective predators and able to grow. Predators also effectively predate the green pus bacteria *Pseudomonas aeruginosa*, and were able to grow. For viruses, *P. erythrophthalma* and *A. hemprichi* are used as predators in experiments, and the predators effectively eliminate the viruses. Rotatoria, with its high filtering ability, is especially effective in eliminating viruses. In this way, predators eat bacteria generally grown based on substrates existing in the sewage water, and also the bacteria that have entered the wastewater. It is well known that protozoa and metazoa eating habits differ greatly depending on the kind of bacteria. *Arthrobactor* and *Micrococcus* are not suitable as food sources, but many colon bacilli become good food sources. By maintaining appropriate numbers of effective protozoa and metazoa in the biological treatment reactor, sanitary safety levels in the treated water is better ensured. Figure 4 shows how protozoa and metazoa eliminate pathogenic bacteria in the biological treatment process.

plant researched		removal rate by activated sludge (%)
'S' plant		
	1	98.8
	2	98.6
	3	98.6
	4	97.6
'K' plant		
	1	99.5
	2	95.7
	3	95.1
	4	98.5

Table 4: Removal of total coliform in sewage treatment process

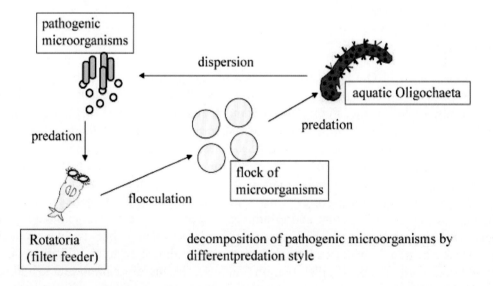

Figure 4: Role of Rotatoria and aquatic Oligochaeta for removal of pathogenic microorganisms

Glossary

AO:	Anaerobic Oxic
A₂O:	Anaerobic Anoxic Oxic
DNA:	Deoxyribo Nucleic Acid
RNA:	Ribo Nucleic Acid
USEPA:	United States Environment Protection Agency
US:	United States

Bibliography

Burkholder J. M. (1999). The lurking perils of Pfiesteria, *Scientific American*, 281(2), 42-49.

Carmichael W. W. (1996). Analysis for microcystins involved in an outbreak of liver failure and death of humans at a hemodialysis center in Caruaru, Pernambuco, Brazil, Proceedings of the 4rd Simposio da sociedade Brasilieira de Toxinologia, Pernambuco, Brazil, October, 85-86.

Hirata T. (1998). Risk of pathogenic protozoa *Cryptosporidium* and *Giardia* in tap water. Manual for risk management of water (ed. E Tsuchiya) [in Japanese], 99-103, Tokyo: Science Forum.

Kaya K. and Watanabe M. M. (1990). Microcystin composition of an axenic clonal strain of *Microcystis viridis* and *Microcystis viridis*-containing waterblooms in Japanese freshwaters, *J. Appl. Phycol.*, 2, 173-178.

Mahakhant A., Sano T., Ratanachot P., Tong-a-ram T., Srivastava V. C., Watanabe M. M. and Kaya K. (1998). Detection of microcystins from cyanobacterial water blooms in Thailand fresh water, *Phycological Research*, 46(suppl.), 25-29.

Rinehart K. L., Namikoshi M. and Choi B. W. (1994). Structural and biosynthesis of toxins from blue-green algae (cyanobacteria), *J. Appl. Phycol.*, 6, 159-176.

Biographical Sketches

Yuhei Inamori is executive researcher of National Institute for Environmental Studies (NIES), where he has been at present post since 1990.Yuhei Inamori received B.S. and M.S. degrees from Kagoshima University, Kagoshima Japan, in 1971 and 1973 respectively. He received Ph. D. degree from Tohoku University, Miyagi, Japan in 1979.

From 1973 to 1979, he was researcher of Meidensha corp. From 1980 to 1984, he was researcher of NIES. From 1985 to 1990, he was senior researcher of NIES. His fields of specification are microbiology, biotechnology and ecological engineering. His current research interests include renovation of water environment using bioengineering and eco-engineering applicable to developing countries. He received Award of Excellent Paper, Japan Sewage Works Association in 1987, Award of Excellent Publication, the Journal "MIZU"in 1991, Award of Excellent Paper, Japanese Society of Water Treatment Biology in 1998.

Naoshi Fujimoto is assistant professor of Tokyo University of Agriculture, where he has been at present post since 2000. Naoshi Fujimoto received B.E. and M.E. degrees from Tohoku University, Miyagi Japan, in 1991 and 1993 respectively. He received D.E. degree in civil engineering from Tohoku University in 1996. From 1996 to 1999, he was research associate of Tokyo University of Agriculture. His field of specification is environmental engineering. His current research interests include control of cyanobacteria and its toxin in lakes and reservoirs.

PHYSICAL/MECHANICAL CONTAMINATION OF WATER

Yuhei Inamori
Executive researcher, National Institute for Environmental Studies, Tsukuba, Japan

Naoshi Fujimoto
Assistant Professor, Faculty of Applied Bioscience, Tokyo University of Agriculture, Tokyo, Japan

Keywords: color, turbidity, odor, taste, pH, alkalinity, hardness, ballast water

Contents

1. Introduction
2. Color and turbidity
3. Odor and taste
4. Alkalinity, pH and Hardness
5. Radionuclides
6. Contamination caused by ships
Glossary
Bibliography
Biographical Sketches

Summary

Problems related to the contamination of water on physical properties are color contamination, offensive odors, corrosiveness (pH), turbidity and radionuclides.

Color contamination of water bodies is caused by metals, dye pollution, soil particles, and by the occurrence of water bloom caused by eutrophication. Color contamination of a water body is a problem because of the harm it inflicts on the scenery and when it used as drinking water or for other domestic uses, its color distresses the users.

Musty odor is a typical offensive odor in drinking water. Microorganisms that produce musty odor are Cyanobacteria, *Oscillatoria*, *Phormidium* and Actinomycetes *Streptomyces*.

Radionuclides may contaminate drinking water source naturally or by human activity. People who use water supply system for drinking, cleaning and showering may be exposed to radiation. International movement of algae species by ballast water may affect the original ecosystem, and may cause domination of algae that produce toxin.

1. Introduction

Physical contamination refers to contamination affecting physical properties of water: coloration, odor, and pH. Such contamination of water does not directly harm people's bodies, but domestic use of the water is very harmful. Problems related to the contamination affecting physical properties are color contamination, offensive odors, corrosiveness (pH), turbidity. Color contamination of water bodies is caused by metals, dye pollution, soil particles, and by the occurrence of water bloom. Offensive odor and taste of water are caused by metals and by microorganisms that produce musty odors. Because residents are distressed if their drinking water and other domestic use water is colored or smells badly, water treatment must completely remove colored or malodorous constituents of water. If the pH value of a water body is far from neutral, it affects the growth of living organisms of the ecosystem, and high acidity of soil or agricultural water prevents the growth of plants and agricultural products. pH is, therefore, an important index of environmental pollution. Contamination of radionuclides is also physical contamination because they emit radiation and affect people's health. Contamination of water by radionuclides was caused by elements such as radon, uranium and radium. Mechanical contamination occurs when exhaust gasses discharged by ships are dissolved by seawater, and when ships spill petroleum. Another cause of contamination is the ballast water that ships carry to remain balanced.

2. Color and Turbidity

Color contamination of water bodies is caused by metals, dye pollution, soil particles, and by the occurrence of water bloom caused by eutrophication. Color contamination of a water body is a problem because of the harm it inflicts on the scenery. And when it used as drinking water or for other domestic uses, its color distresses the users. Typical algae species that color water resources such as rivers, lakes and reservoirs include Cyanobacteria (blue-green algae) *Microcystis, Oscillatoria, Phormidium, Anabaena,* and *Aphanizomenon,* Chlorophyceae (green algae) *Euglena, Closterium,* and *Ankistrodesmus,* Bacillariophyceae *Synedra* and *Melosira,* and the Flagellataes, *Peridiuium,* and others. When Cyanobacteria and Chlorophyceae form a bloom, the water body is colored by their pigments. When Bacillariophyceae have formed a bloom, the water turns brown. The Flagellatae *Peridiuium* forms bloom in dam reservoirs, turning its water reddish-brown.

Color pollution is also caused by the metals: aluminum, copper, iron, and manganese. When algae die, they sink to the bottom and are decomposed by bacteria. Dissolved oxygen of the bottom layer decreases, leaching out the iron or manganese from sediment occurs. If the water resource is contaminated by iron or manganese, laundry is colored and the taste of the water is harmed. It is, therefore, essential that water treatment include measures to remove iron and manganese. Objective values for the metals that color water specified under water quality in Waterworks Law (Japan) include those for manganese, aluminum, and copper, and the Guidelines for Drinking Water Quality of WHO stipulates objective values for the aluminum, copper, iron, and manganese (Table 1).

Color contamination is caused by the inflow of wastewater containing dye. Dyes include direct dyes, acid dyes, and basic dyes, and dye wastewater discharged by dye factories contains large quantities of inorganic salts, organic acids, surfactant, water-soluble polymer compounds, and so on. Dye wastewater not only causes color contamination; other problems are that many dyes contain persistent substances and some are toxic.

Water quality in Waterworks Law in Japan	Manganese	$\leqq 0.05 mg \cdot l^{-1}$
	Aluminum	$\leqq 0.2 mg \cdot l^{-1}$
	Copper	$\leqq 1 mg \cdot l^{-1}$
Objective properties as drinking water in Guidelines for Drinking Water Quality of WHO	Aluminum	$0.2 mg \cdot l^{-1}$
	Copper	$1 mg \cdot l^{-1}$
	Iron	$0.3 mg \cdot l^{-1}$
	Manganese	$0.1 mg \cdot l^{-1}$

Table 1. Standards for chemicals causing color in drinking water

Humic acid and humin are coloring agents formed by the transformation of the remains of plants and animals into dark or blackish brown humus by the action of microorganisms in soil. Lakes contaminated by large quantities of these substances are called dystrophic lakes, and their water is either yellowish brown or brown. Because humin is a color contamination agent and also reacts with chlorine to form the carcinogen, trihalomethane, when a large concentration of humin is in water resource, the humin must be completely removed. Environmental water standards concerning color are not established, but objective value for drinking water is set at 15 TCU in the Guidelines for Drinking Water Quality of the WHO.

After rain falls, brown turbidity appears in rivers, reservoirs, lakes, and marshes as a result of rainwater carrying soil into their waters. Water bodies in regions where sediment with low specific gravity called laterite has been deposited are always turbid because of sediment. Turbidity seriously harms the scenery and detracts from the hydrophilic role of water bodies by, for example, discouraging swimming. When turbid water is obtained as a source of drinking water, it damages treatment plants by plugging their treatment systems. As turbidity indices, environmental standards set suspended solid levels measured using glass fiber filter paper. In the Guidelines for Drinking Water Quality of the WHO, the objective value for potable water is 5 NTU.

3. Odor and Taste

Musty odor is a typical offensive odor in drinking water. Microorganisms that produce musty odor are the Cyanobacteria, *Oscillatoria*, *Phormidium* and the Actinomycetes *Streptomyces*. The principal musty substances are geosmin, and 2-methylisoborneol, and the limit values for the detection of their odor are 0.01 $\mu g \bullet l^{-1}$ for geosmin and 0.007 $\mu g \bullet l^{-1}$ for 2-methylisoborneol. Effective ways to prevent the production of musty odors are removal of nutrient such as nitrogen and phosphorus and aerating lake water, but it can also be removed by slow filtration, activated carbon treatment, and ozone treatment. Items relating to comfortable water quality in Waterworks Law in Japan include standards for geosmin and 2-methylisoborneol (Table 2), and their concentrations at permanent treatment facilities using granular activated carbon etc. are both less than 0.00001 $mg \bullet l^{-1}$.

In addition to musty odor, objective values are also set for odor intensity and free carbon dioxide in the items relating comfortable water quality. As offensive odor measures, the Guidelines for Drinking Water Quality of the WHO sets objective values for the inorganic substances such as hydrogen sulfide and ammonia, and for the organic substances such as toluene and xylene (Table 2). These malodorous substances are basically produced by artificial pollution, but there are cases where hydrogen sulfide and ammonia are produced naturally.

Water quality in Waterworks Law in Japan	2-Methylisoborneol	$\leqq 0.00001$ mg $\cdot$ l^{-1}
	Geosmin	$\leqq 0.00001$ mg $\cdot$ l^{-1}
	Chloride	$\leqq 200$ mg $\cdot$ l^{-1}
	Hardness (Calcium, Magnesium)	$\leqq 300$ mg $\cdot$ l^{-1}
	Iron	$\leqq 0.3$ mg $\cdot$ l^{-1}
	Sodium	$\leqq 200$ mg $\cdot$ l^{-1}
	Phenols	$\leqq 0.005$ mg $\cdot$ l^{-1}
Objective properties as drinking water in the Guidelines for Drinking Water Quality of the WHO	Ammonia	1.5 mg $\cdot$ l^{-1}
	Chloride	250 mg $\cdot$ l^{-1}
	Hydrogen sulfide	0.05 mg $\cdot$ l^{-1}
	Sodium	200 mg $\cdot$ l^{-1}
	Sulfate	250 mg $\cdot$ l^{-1}
	Dissolved compounds	1000 mg $\cdot$ l^{-1}
	Zinc	3 mg $\cdot$ l^{-1}
	Toluene	24-170 ug $\cdot$ l^{-1}
	Xylene	20-1800 ug $\cdot$ l^{-1}
	Ethylbenzene	2-200 ug $\cdot$ l^{-1}
	Styrene	4-2600 ug $\cdot$ l^{-1}
	Monochlorobenzene	10-120 ug $\cdot$ l^{-1}
	1,2-Dichlorobenzene	1-10 ug $\cdot$ l^{-1}
	1,4-Dichlorobenzene	0.3-30 ug $\cdot$ l^{-1}
	Trichlorobenzene	5-50 ug $\cdot$ l^{-1}
	Surfactant	-
	Chlorine	0.6-1.0 ug $\cdot$ l^{-1}
	2-Chlorophenol	0.1-10 ug $\cdot$ l^{-1}
	2,4-Dichlorophenol	0.3-40 ug $\cdot$ l^{-1}
	2,4,6-Trichlorophenol	2-300 ug $\cdot$ l^{-1}

Table 2. Standards for substances relating offensive odor and taste in drinking water

4. Alkalinity, pH and Hardness

Factors governing the pH of water resources include acid rain and seepage from landfill. If alkalinity is low, or in other words, if the acidity is high, pipes are harmed, and pollution may be caused by heavy metals in pipes. USEPA recommends groundwater alkalinity of 60 mg•l^{-1}. According to the Guidelines for Drinking Water Quality of the WHO, low pH is putrefactive, and high pH of drinking water harms its taste and feel. A pH value of 8 or less is desirable to obtain sterilization effect. The water quality standard of Japan's Waterworks Law sets a standard pH range of 5.6 to 8.6.

When hard water is used, boiler scale forms in utensils and water heaters, and soap does not foam. Causes of hard water include calcium and potassium flowing into water from soil or aquifers that contain limestone or dolomite. For these reasons, the water quality standards of the Japanese Waterworks Law stipulate that hardness must be 300 mg•1^{-1} or less. $Ca(OH)_2$ or Na_2CO_3 are used to remove hardness.

5. Radionuclides

Radionuclides are known to have effects on health of people. If water source for drinking water contains radionuclides over the concentration harmful to health, removal of radionuclides is necessary. Radionuclides are found in air, water and soil. People are exposed to background levels of radiation all the time. Annual average of natural radiation in the world is estimated 2.4 mCv per capita. Radionuclides are found as trace elements in most rocks and solid and are formed principally by the radioactive decay of uranium-238 and thorium-232, which are the long-lived parent elements of the decay series. Decay occurs by the emission of an alpha particle or a beta particle and gamma rays from the nucleus of the radioactive element.

Radionuclides generally dissolve drinking water source through the erosion or chemical weathering of naturally occurring mineral deposits, and human activity such as mining and industrial activities. People who use water supply system treating ground water for drinking, cleaning and showering may be exposed to radiation. Ground water flows through rocks which contain uranium and radium which release radon in their decay series. The most common radionuclides in ground water are radon-222, radium-226, uranium-238, and uranium-234 of the uranium-238 decay series, and radium-228 of the thorium-232 decay series. Concentration of radionuclides such as uranium, radium and radon in ground water is higher than that of surface water such as lakes, reservoirs and rivers.

Evidence suggests that long-term exposure to radionuclides in drinking water may cause cancer. Chemical reaction in cell is affected by radiation of radionuclides. Radiation pulls electrons off atoms in the cells and may prevent function of the cell, leading cell's death, inability of repair and uncontrolled growth (cancer). Radon is supposed most important radionuclides for its health effects. Radon can enter the body by direct water consumption, or through inhalation when radon is liberated from water used for cleanimg, showering and other purposes. Radon decays and its charge alpha emitting progeny attach to dust, cigarette smoke, and other aerosol particles. These particles are inhaled to the lung interior and cells are exposed by radiation. Thus it is suggested that radon may cause lung cancer.

EPA has limits of radionuclides in drinking water called maximum contaminant levels (MCLs) as beta/photon emitters 4mrem/year, gross alpha particle 15 pCi/L, radium-226 and radium-228 5pCi/L and uranium 30 ug/L. Gross alpha and beta activities is set at 0.1Bq/L and 1Bq/L in the Guidelines for Drinking Water Quality of the WHO. In drinking water systems, physical or chemical treatment that remove radionuclides from ground water are installed. Radon is removed by granular activated carbon process and aeration process. Radium and uranium are removed by treatment processes such as ion-exchange, reverse osmosis, electro-dialysis, coagulation and lime softening.

6. Contamination Caused by Ships

Ballast water is sea water carried inside large cargo vessels, tankers, and other ships to guarantee their safety. A cargo ship takes sea water into its hull to stabilize it when its weight has declined after it has unloaded its cargo at its destination. After it returns to its home port it discharges the ballast water as it loads new cargo. But the sea water taken into the ship contains microorganisms including harmful algae that inhabit the ocean waters. It is reported that shellfish eat the harmful algae that grows there and people who eat these shellfish suffer food poisoning. There are even reports of deaths caused by this process. There is also a danger of discharged ballast water bringing in foreign species disrupting ecosystems. This contamination is caused by the international movement of algae species that cause red tide, or algae species that produce toxins, and although it is classified as biological contamination, essentially it can be viewed as physical contamination caused by the movement of sea water.

A famous case of the effects of cross-border movement of ballast water is the harmful plankton the Flagellatae *Alexandrium* carried from Japan to Australia, where it caused serious problems by producing the deadly poison called saxitoxin. The *Alexandrium* has been designated as a biological weapon, and it is well-known as a type of algae whose international movement is strictly controlled. At this time, there are countries that strictly control the transport of ballast water from other regions, but in some countries, adequate protective measures have not been taken. Standards are now being established to develop practical technology so that ship owners can be required to biologically, physically, and chemically treat ballast water. One ballast water measure now carried out is replacing the ballast water far out at sea (re-ballasting). But this method has shortcomings. Considering the safety of ship's hulls and marine and meteorological conditions, it is a difficult method to implement. It is also difficult to use on short shipping routes and it increases the work the crew must perform. It is, therefore, difficult to describe it as a perfect solution. Methods that can replace re-ballasting are extremely important. Research is now being carried out on a variety of measures to replace re-ballasting. They include temperature control, ultraviolet irradiation, electrical treatment, filtration, sonic methods, magnetic methods, and chemical treatment methods.

About 50% of all marine contamination is caused by oil, and this can be called one of the most serious of all environmental contamination problems. Oil contamination is mechanical contamination caused by tanker accidents, the illegal discharge of contaminated water from the bottom of ships and of water used to clean tankers, oil-field accidents, and the discharge of ballast. Because such contamination by oil affects the ecosystems of aquatic life, it is essential to accurately clarify the state of marine environments and undertake suitable measures.

Glossary

NTU:	Nephelometric Turbidity Unit
TCU:	True Color Unit
USEPA:	United States Environmental Protection Agency
WHO:	World Health Organization

Bibliography

United States Environmental Protection Agency Office of Ground Water and Drinking Water (2002) Radionuclides in Drinking Water: A Small Entity Compliance Guide, EPA 815-R-02-001.[This is a guide of radionuclides in drinking water for operating community water system].

UNSCEAR (1988). Sources, effects and risks of ionizing radiation. 1988 report to the general assembly with annexes. New York, United Nations.[This is a report on sources, effects and risks of natural and artificial radiation].

World Health Organization (1994). Guidelines for drinking-water quality Vol. 1 Recomendation, WHO, Geneva.[This presents property of drinking water from microbiological, biological, chemical-physical aspects].World Health Organization (1994). Guideliness for drinking-water quality Vol. 1 Recomendation, WHO, Geneva.

World Health Organization (1994). Guidelines for drinking-water quality Vol. 2 Health criteria and other supporting information, WHO, Geneva.[This presents supporting information of guideline value such as animal assays, safety coefficient, risk assessment of carcinogenic substances].World Health Organization (1994). Guideliness for drinking-water quality Vol. 2 Health criteria and other supporting information, WHO, Geneva.

Zapecza O. S. and Szabo Z. (1986). Natural radioactivity in ground water-a review, National Water Summary 1986-Ground Water Quality, 50-57.[This is a review on radionuclides in ground water: geochemistry, distribution, standards, health effects and removal from ground water].

Biographical Sketches

Yuhei Inamori is executive researcher of National Institute for Environmental Studies (NIES), where he has been at present post since 1990.Yuhei Inamori received B.S. and M.S. degrees from Kagoshima University, Kagoshima Japan, in 1971 and 1973 respectively. He received Ph. D. degree from Tohoku University, Miyagi, Japan in 1979.

From 1973 to 1979, he was researcher of Meidensha corp. From 1980 to 1984, he was researcher of NIES. From 1985 to 1990, he was senior researcher of NIES. His fields of specification are microbiology, biotechnology and ecological engineering. His current research interests include renovation of water environment using bioengineering and eco-engineering applicable to developing countries. He received Award of Excellent Paper, Japan Sewage Works Association in 1987, Award of Excellent Publication, the Journal "MIZU"in 1991, Award of Excellent Paper, Japanese Society of Water Treatment Biology in 1998.

Naoshi Fujimoto is assistant professor of Tokyo University of Agriculture, where he has been at present post since 2000. Naoshi Fujimoto received B.E. and M.E. degrees from Tohoku University, Miyagi Japan, in 1991 and 1993 respectively. He received D.E. degree in civil engineering from Tohoku University in 1996. From 1996 to 1999, he was research associate of Tokyo University of Agriculture. His field of specification is environmental engineering. His current research interests include control of cyanobacteria and its toxin in lakes and reservoirs.

INDEX

A

Agricultural
    Chemicals, 21, 29, 36-41, 51-54, 56, 62, 81, 85, 101-102, 109-110, 127, 131-132
        Waste, 156-157
Alkalinity, pH And Hardness, 188, 192
Aluminum, 43-44, 50, 57-58, 78, 86, 150, 156, 161, 170, 175, 189
Alzheimer Disease, 156, 170, 174
Anabaena Macrospora, 140, 142
Anatoxin-a(s), 140, 142, 153
Antimony, 156-157, 162-163, 176
Aphanizomenon Flos-Aquae, 140, 142, 153, 182
Aphantoxin, 140, 142, 153, 182
Arsenic, 21-24, 33-34, 48-50, 54, 132-133, 156-159, 162, 176

B

Ballast Water, 137-138, 185, 188-189, 193
Barium, 156-157, 162-163, 175-176
BHC, 140, 147, 149, 153
Biological Treatment, 46-47, 64-65, 67, 178-179, 183, 185-186
Blue-Green Algae, 41, 139-142, 155, 187, 189
Boron, 33-34, 55, 99, 156-161, 175-176

C

Cadmium, 2, 21-23, 32, 48, 50, 54, 132-133, 156-157, 162-164, 174-175
Calcium, 68, 89, 92-95, 99, 122, 129, 191-192
Causative Materials
    Contaminating Groundwater, 22
    For Contamination of Surface Water, 36-37
Causes of Salinization and Saline Soil, 88-89
Characteristics of Organic Pollutants in Each Water Area, 140, 153
Chemical
    Management, 1
    Substances, 3, 21-22, 32, 36, 89, 101-102, 109, 133-135, 137, 144
Chernobyl, 156, 173-174
Chlorinated Organic Compounds, 21, 26, 36
Chromium, 21-23, 33, 48-51, 55, 62, 67, 132, 156-157, 162, 164-165, 176
Co-PCB, 140, 150, 153-154
Color and Turbidity, 188-189
Combined Sewage System, 122
Contamination
    By
        Hazardous
            Microorganisms, 131, 136
            Substances, 131-132
        Industrial Chemicals, 140, 143
    Of
        Metabolites Produced By Aquatic Microorganisms, 140-141
        Water Resources, 131-139
Copper, 1, 44, 50, 55, 156-157, 162, 170-171, 176, 189
Countermeasures
    Against Water Pollution, 40, 53
    For
        Point Sources, 61

ENCYCLOPEDIA OF LIFE SUPPORT SYSTEMS (EOLSS)

A source of knowledge for sustainable development and global security to lead to fulfillment of human needs through simultaneous socio-economic and technological progress and conservation of the Earth's natural systems

It is a virtual dynamic library with contributions from thousands of scholars from over 100 countries and edited by well over 300 subject experts, for a wide audience: pre-university/university students, educators, professional practitioners, informed specialists, researchers, policy analysts, managers, and decision makers.

This archive is now available at **http://www.eolss.net** as an integrated compendium of twenty component encyclopedias

1. Earth and Atmospheric Sciences
2. Mathematical Sciences
3. Biological, Physiological and Health Sciences
4. Biotechnology
5. Land Use, Land Cover and Soil Sciences
6. Tropical Biology and Conservation
7. Social Sciences and Humanities
8. Physical Sciences, Engineering and Technology Resources
9. Control Systems, Robotics and Automation
10. Chemical Sciences Engineering and Technology Resources
11. Water Sciences, Engineering and Technology Resources
12. Energy Sciences, Engineering and Technology Resources
13. Environmental and Ecological Sciences, Engineering and Technology Resources
14. Food and Agricultural Sciences, Engineering and Technology Resources
15. Human Resources Policy, Development and Management
16. Natural Resources Policy and Management
17. Development and Economic Sciences
18. Institutional and Infrastructural Resources
19. Technology, Information and System Management Resources
20. Area Studies (Africa, Brazil, Canada and USA, China, Europe, Japan, Russia)

Within these twenty on-line encyclopedias, there are about 235 Themes, each of which has been compiled under the editorial supervision of a recognized world expert or a team of experts such as an International Commission specially appointed for the purpose. Each of these 'Honorary Theme Editors' was responsible for selection and appointment of authors to produce the material specified by EOLSS. On average each Theme contains about thirty chapters. It deals in detail with *interdisciplinary* subjects, but it is also *disciplinary*, as each major core subject is covered in great depth by world experts.

EOLSS Presentations are in gradually increasing depth.

- **Level 1 (Theme Level)**: Presentations of broad perspectives of major subjects

- **Level 2 (Topic Level)**: Presentations of perspectives of the special topics within the subjects
- **Level 3 (Article Level)**: In-depth presentations of the subjects in various aspects

Some themes have a Three level structure and some Two level structure. The EOLSS web of knowledge is woven over a hierarchical structure through cross reference links

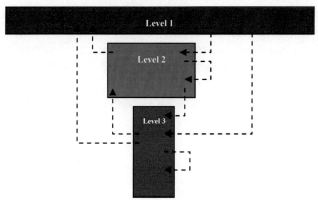

The three level structure

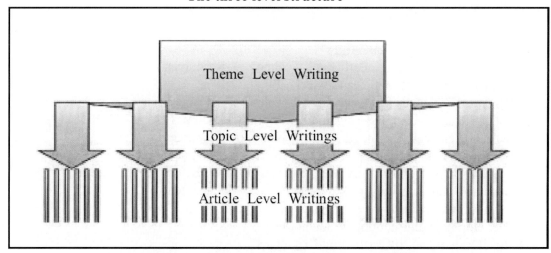

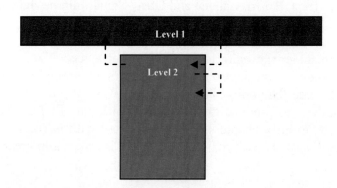

The two level structure

EOLSS is the result of an unprecedented global effort involving the collective expertise of nearly 400 Subject Editors and 10000Authors from over 100 countries.

The EOLSS project is coordinated by the UNESCO-EOLSS Joint Committee (http://www.eolss.net/eolss_international.aspx) and sponsored by Eolss Publishers, which is based in Oxford, United Kingdom. Through many and diverse consultation meetings around the world, the EOLSS has benefited immensely from the academic, intellectual, and scholarly advice of each and every member of a large International Editorial Council (http://www.eolss.net/eolss_international.aspx), which includes Nobel and UN Kalinga Laureates, World Food Prize Laureates, and several fellows of academies of science and engineering of countries throughout the world.

From 1996 thousands of scientists, engineers and policy-makers began meeting just to define the scope of the project, before discussing the details of the contributions. Regional workshops were held in Washington DC, Tokyo, Moscow, Mexico City, Beijing, Panama, Abu Sultan (Egypt), and Kuala Lumpur, to develop a list of possible subjects and debate analytical approaches for treating them.

A UNESCO-EOLSS Joint Committee was established with the objectives of (a) seeking, selecting, inviting and appointing Honorary Theme Editors (HTEs)/International Commissions for the Themes, (b) providing assistance to them, (c) obtaining appropriate contributions for the different levels of the Encyclopedia, and (d) monitoring the text development (See UNESCO-EOLSS Joint Committee below).

Motivation for Publication in Electronic Media

Print projects for such large publications are highly material and energy intensive, a consideration that is very important in the case of the EOLSS since it advocates minimal use of natural resources.

Another great advantage of the electronic medium for large publications, in addition to portability, is the search facility. For instance, information on any chosen aspect can be found by search with the click of a button from the huge body of knowledge. Parts of the information may be downloaded, organized, printed, or adapted to one's own academic needs such as research studies, course-packs, etc. This is not the case with the print versions.

EOLSS Updating and Augmentation

Efforts to update and augment the EOLSS body of knowledge: Editors and Authors are given free access to access their works and forward updates if any. New themes are added covering many additional subjects. Updating/augmenting is frequent- almost every month

EOLSS in Official UN Languages

The Universal Networking Digital Language Foundation (UNDL) is now active on the translation project. The 1st International Workshop on Natural Language Processing Using the Universal Networking Language (UNL), took place during 4-7 May 2007 at Alexandria, Egypt. UNDL started to work with the six official UN languages. Twelve

UNL language centers situated in France, Russia USA, Cairo, Sao Paulo, Madrid, Tokyo, China, etc. are participating in this work.

Unite Nations DECADE OF EDUCATION FOR SUSTAINABLE DEVELOPMENT: 2005-2014

EOLSS is dedicated to contribute to the implementation of Unite Nations DESD: 2005-2014

A sample of opinions on the EOLSS project

"Scientific curiosity and technological endeavors are deeply rooted aspects of human nature. They are responsible for the development of human society and welfare. But they are also responsible for much of the environmental problems we are facing today. Solutions for these shortcomings are inconceivable without full scientific and technological support. EOLSS has the goal to provide a firm knowledge base for future activities to prolong the lifetime of the human race in a hospitable environment." **Richard R. Ernst, Nobel Laureate- Chemistry**

"The EOLSS is not only appropriate, but it is imaginative and, to my knowledge, unique. Much of what we can write about science, about energy, about our far-ranging knowledge base, can indeed be found in major encyclopedias, but as I understand your vision, never as a central theme; the theme of humanity, embedded in nature and constrained to find ways of maintaining a relationship with nature based upon understanding and respect." **Leon M. Lederman, Nobel Laureate in Physics**

"The population of our planet and its development over the ages sets the scene for considering all global problems and it is reasonable to begin their discussion with population growth. ... Thus we are dealing with an interdisciplinary problem in an attempt to describe the total human experience, right from its very beginning. But without this perspective of time it is not possible to objectively assess what is happening today and provide an objective view of the present state of development, the challenge now facing humanity." **S.P. Kapitza, UNESCO Kalinga Prize Winner**

"Pursuit of knowledge and truth supersedes present considerations of what nature, life or the world are or should be, for our own vision can only be a narrow one. Ethical evaluation and rules of justice have changed and will change over time and will have to adapt. Law is made for man, not man for law. If it does not fit any more, change it.... Some think that it is being arrogant to try to modify nature; arrogance is to claim that we are perfect as we are! With all the caution that must be exercised and despite the risks that will be encountered, carefully pondering each step, mankind must and will continue along its path, for we have no right to switch off the lights of the future.... We have to walk the path from the tree of knowledge to the control of destiny." **Jean-Marie Lehn, Nobel Laureate in Chemistry**

"Ecotechnology involving appropriate blends of traditional technologies and the ecological prudence of the past with frontier technologies such as biotechnology, information technology, space technology, new materials, renewable energy technology and management technology, can help us to promote global sustainable development involving harmony between humankind and nature on the one hand and tolerance and love of diversity and pluralism in human societies on the other. We need shifts in technology and public policy. This is a challenging task to which the Encyclopedia of Life Support Systems should address itself." **M.S. Swaminathan, First World Food Prize winner**

"EOLSS is concerned with the Life Support Systems... Each of these systems is a very complex one. ...we have to think of all these "systems" as closely related "subsystems" of the Planet Earth System. ...Rational decisions will be more and more possible to envision if one will be able to couple the physical modeling to economic and financial models and to human factors..." **J.L. Lions, Japan Prize winner in Applied Mathematics**

"In the coming decades, the need for working together, across countries, is essential. We need to bring 3 billion poor into society and make it possible for them to lead healthy lives, to get the health interventions that we all take as something as self-evident. It can be done. It is within reach. We need to invest in the future. We need to invest in the health of all people worldwide." **Gro Harlem Brundtland, Director General, WHO,** in an interview with Patricia Morales

"We must learn to live at one with nature. Nature does not bear grudges, but it must not be brought to the point where it can no longer sustain human society and the continuance of humankind on Earth. I believe that one of the most important things is the shaping of a new value system, because nature can live without us, but we cannot live without nature. Instead of a hedonic approach, we should promote an approach that reasonably limits consumerism and which promotes the virtue of enoughness. If we insist on consumerism as the new utopia, nature will reject such a system, as surely as cultural diversity rejected the totalitarian system. Our generation has to face a difficult challenge, but as recent history has proven, walls of difficulty, like the Berlin Wall, can fall." **Mikhail Gorbachev, Nobel Laureate in Peace,** in an interview with Patricia Morales

"We live in an increasingly interdependent world. As a result, we must cooperate together across all boundaries of nation, culture, faith, and race, and at all levels—locally, nationally, regionally, and globally— if we are to achieve our basic environmental, economic and social goals. Furthermore, if we are to make wise choices and to cooperate together effectively, we urgently need a shared vision of fundamental ethical values to guide us. In other words, the development of global ethics is essential. Our very survival as a species is in doubt if we cannot clarify our ethics and develop common values around such basic issues as environmental protection, justice, human rights, cultural diversity, economic equity, eradication of poverty, and peace." **Steven C. Rockefeller, Earth Charter Commissioner,** in an interview with Patricia Morales

EOLSS e-Books and their full color print versions in subject categories

For the convenience of users' special interests the EOLSS body of knowledge is being made available in e-book (pdf) format with the provision for Print on Demand. There is a wide gallery of e-books making up a virtual library and the sets of volumes classified under different component encyclopedias are outlined below. **The ISBN numbers are respectively eISBN for e-books and ISBN for print versions. Forthcoming items are without these numbers.**

Prices of e-Books and their full color print versions

Each e-book (Adobe Reader-single user) volume: **US$78/Euros60/UKP50**
Each print volume (Full Color Library Edition): **US$234/Euros180/UKP150**
Orders from the USA, Europe, and UK will be charged in US$, Euros, UK Pounds respectively.
Orders from all other regions will be charged in US$.
e-Book orders for multiple users will be entitled to get a discount: Five users-10%; Ten users-20%
To place your order visit **http://www.eolss.net/**

DESWARE: Encyclopedia of Desalination and Water Resources

This is a unique publication, the only one of its kind in the filed of desalination and water resources which puts together contributions from leaders from all over the world. For details visit (**www.desware.net**)

EARTH AND ATMOSPHERIC SCIENCES

**Environmental Structure and Function: Earth System

Editors: Nikita Glazovsky, *Institute of Geography, Russian Academy of Sciences, Russia*

Nina Zaitseva, *Department of Earth Sciences, Russian Academy of Sciences, Moscow, Russia*
eISBN:
ISBN:

Environmental Structure and Function: Climate System (Vols.1-2)

Editor: George Vadimovich Gruza, *Institute of Global Climate and Ecology, Russian Academy of Sciences, Russia*
eISBN: 978-1-84826-288-1, 978-1-84826-289-8
ISBN: 978-1-84826-738-1, 978-1-84826-739-8

Natural Disasters (Vols. 1-2)

Editor: Vladimir M. Kotlyakov, *Institute of Geography, Russian Academy of Sciences, Russia*
eISBN: 978-1-84826-309-3, 978-1-84826-310-9
ISBN: 978-1-84826-759-6, 978-1-84826-760-2

Geography (Vols. 1-2)

Editor: Maria Sala, *University of Barcelona, Spain*
eISBN: 978-1-905839-60-5, 978-1-905839-61-2
ISBN: 978-1-84826-960-6, 978-1-84826-961-3

Geology (Vols. 1-5)

Editors: Benedetto De Vivo, *Universita di Napoli "Federico II", Italy*

Bernhard Grasemann, *University of Vienna, Austria*

Kurt Stüwe, *Institut für Geologie und Paläontologie, Universitaet Graz, Austria*
eISBN: 978-1-84826-004-7, 978-1-84826-005-4, 978-1-84826-006-1, 978-1-84826-007-8, 978-1-84826-008-5
ISBN: 978-1-84826-454-0, 978-1-84826-455-7, 978-1-84826-456-4, 978-1-84826-457-1, 978-1-84826-458-8

Geophysics and Geochemistry (Vols. 1-3)

Editor: Jan Lastovicka, *Institute of Atmospheric Physics, Academy of Sciences of the Czech Republic, Czech Republic*
eISBN: 978-1-84826-245-4, 978-1-84826-246-1, 978-1-84826-247-8
ISBN: 978-1-84826-636-0, 978-1-84826-656-8, 978-1-84826-662-9

Oceanography (Vols. 1-3)

Editors: Jacques C.J. Nihoul, *University of Liege, Belgium*

Chen-Tung Arthur Chen, *National Sun Yat-Sen University, Taiwan*
eISBN: 978-1-90583-962-9, 978-1-90583-963-6, 978-1-90583-964-3
ISBN: 978-1-84826-962-0, 978-1-84826-963-7, 978-1-84826-964-4

Geoinformatics (Vols. 1-2)

Editor: Peter M. Atkinson, *University of Southampton, UK*
eISBN: 978-1-905839-87-2, 978-1-905839-86-5

ISBN: 978-1-84826-987-3, 978-1-84826-986-6

Advanced Geographic Information Systems (Vols. 1-2)

Editor: Claudia Maria Bauzer Medeiros, *Universidade Estadual de Campinas (UNICAMP), Brazil*
eISBN: 978-1-905839-91-9, 978-1-905839-92-6
ISBN: 978-1-84826-991-0, 978-1-84826-992-7

**Astronomy and Astrophysics

Editors: Oddbjørn Engvold, *University of Oslo, Norway*

Rolf Stabell, *University of Oslo, Norway*

Bozena Czerny, *Copernicus Astronomical Center, Bartycka 18, 00-716 Warsaw, Poland*

John Lattanzio, *Monash University, Victoria Australia*
eISBN:
ISBN:

**Tropical Meteorology

Editor: Yuqing Wang, *Department of Meteorology and International Pacific Research Center, School of Ocean and Earth Science and Technology University of Hawaii at Manoa*
eISBN:
ISBN:

MATHEMATICAL SCIENCES

Environmetrics

Editors: Abdel H. El-Shaarawi, *National Water research Institute, Burlington Ontario Canada,*

Jana Jureckova *Charles University Sokolovska, Czech Republic*
eISBN: 978-1-84826-102-0
ISBN: 978-1-84826-552-3

Biometrics (Vols. 1-2)

Editors: Susan R. Wilson, *School of Mathematical Sciences, Australian National University, Australia*

Conrad Burden, *The Australian National University, Australia*
eISBN: 978-1-905839-36-0, 978-1-905839-37-7
ISBN: 978-1-84826-936-1, 978-1-84826-937-8

Mathematics: Concepts, and Foundations (Vols. 1-3)

Editor: Huzihiro Araki, *Research Institute of Mathematical Sciences, Kyoto University, Japan*
eISBN: 978-1-84826-130-3, 978-1-84826-131-0, 978-1-84826-132-7
ISBN: 978-1-84826-580-6, 978-1-84826-581-3, 978-1-84826-582-0

Probability and Statistics (Vols. 1-3)

Editor: Reinhard Viertl, *Technische Universität Wien, Austria*
eISBN: 978-1-84826-052-8, 978-1-84826-053-5, 978-1-84826-054-2
ISBN: 978-1-84826-502-8, 978-1-84826-503-5, 978-1-84826-504-2

Mathematical Models of Life Support Systems (Vols. 1-2)

Editors: Valeri I. Agoshkov, *Institute of Numerical Mathematics, Russian Academy of Sciences, Russia*

Jean-Pierre Puel, *Universitetde Versailles St. Quentin, Versailles, France*
eISBN: 978-1-84826-128-0, 978-1-84826-129-7
ISBN: 978-1-84826-578-3, 978-1-84826-579-0

Mathematical Models (Vols. 1-3)

Editor: Jerzy A. Filar, *University of South Australia, Australia*

Jacek B. Krawczyk, *Victoria University of Wellington, New Zealand*
eISBN: 978-1-84826-242-3, 978-1-84826-243-0, 978-1-84826-244-7
ISBN: 978-1-84826-695-7, 978-1-84826-696-4, 978-1-84826-697-1

Computational Models (Vols. 1-2)

Editors: Vladimir V. Shaidurov, *Institute of Computational Modeling of the Russian Academy of Sciences, Russia*

Olivier Pironneau, *Universite de Paris 6, UFR France*
eISBN: 978-1-84826-035-1, 978-1-84826-036-8
ISBN: 978-1-84826-485-4, 978-1-84826-486-1

Optimization and Operations Research (Vols. 1-4)

Editor: Ulrich Derigs, *University of Cologne, Germany*
eISBN: 978-1-905839-48-3, 978-1-905839-49-0, 978-1-905839-50-6, 978-1-905839-51-3
ISBN: 978-1-84826-948-4, 978-1-84826-949-1, 978-1-84826-950-7, 978-1-84826-951-4

Systems Science and Cybernetics (Vols. 1-3)

Editor: Francisco Parra-Luna, *Universidad Complutense de Madrid, Spain*
eISBN: 978-1-84826-202-7, 978-1-84826-203-4, 978-1-84826-204-1
ISBN: 978-1-84826-652-0, 978-1-84826-653-7, 978-1-84826-654-4

**History of Mathematics

Editors: Vagn Lundsgaard Hansen, *Technical University of Denmark, Denmark*
Jeremy Gray, *Centre for Mathematical Sciences, Open University, UK*
eISBN:
ISBN:

Mathematical Models in Economics (Vols. 1-2)

Editor: Wei-Bin Zhang, *Ritsumeikan Asia Pacific University, Japan*
eISBN: 978-1-84826-228-7, 978-1-84826-229-4
ISBN: 978-1-84826-678-0, 978-1-84826-679-7

**Mathematical Physiology

Editor: Andrea de Gaetano, *CNR IASI Laboratorio di Biomatematica, Roma, Italy*
eISBN:
ISBN:

BIOLOGICAL, PHYSIOLOGICAL AND HEALTH SCIENCES

Global Perspectives in Health (Vols. 1-2)

Editor: B. P. Mansourian, *Research Policy & Cooperation, WHO, Switzerland*
eISBN: 978-1-84826-248-5, 978-1-84826-249-2
ISBN: 978-1-84826-698-8, 978-1-84826-699-5

Water and Health (Vols. 1-2)

Editor: W.O.K. Grabow, *University of Pretoria, South Africa*
eISBN: 978-1-84826-182-2, 978-1-84826-183-9
ISBN: 978-1-84826-632-2, 978-1-84826-633-9

Physical (Biological) Anthropology

Editor: P. Rudan, *Institute for Anthropological Research, University of Zagreb, Zagreb, Croatia*
eISBN: 978-1-84826-226-3
ISBN: 978-1-84826-676-6

Psychology (Vols. 1-3)

Editor: Stefano Carta, *University of Cagliari, Italy*
eISBN: 978-1-905839-65-0, 978-1-905839-66-7, 978-1-905839-67-4
ISBN: 978-1-84826-965-1, 978-1-84826-966-8, 978-1-84826-967-5

Genetics and Molecular Biology

Editor: Kohji Hasunuma, *Yokohama City University, Japan*
eISBN: 978-1-84826-123-5
ISBN: 978-1-84826-573-8

Physiology and Maintenance (Vols. 1-5)

Editors: Osmo Otto Päiviö Hänninen and Mustafa Atalay, *University of Kuopio, Finland*

Volume 1: Physiology and Maintenance
General Physiology
eISBN- 978-1-84826-039-9
ISBN- 978-1-84826-489-2

Volume 2: Physiology and Maintenance
Enzymes: The Biological Catalysts of Life Nutrition and Digestion
eISBN- 978-1-84826-040-5
ISBN- 978-1-84826-490-8

Volume 3: Physiology and Maintenance
Renal Excretion, Endocrinology, Respiration
Blood Circulation: Its Dynamics and Physiological Control
eISBN- 978-1-84826-041-2
ISBN- 978-1-84826-491-5

Volume 4: Physiology and Maintenance
Locomotion in Sedentary Societies
eISBN- 978-1-84826-042-9
ISBN- 978-1-84826-492-2

Volume 5: Physiology and Maintenance
Neurophysiology
Plant Physiology and Environment: A Synopsis
eISBN- 978-1-84826-043-6
ISBN- 978-1-84826-493-9

Medical Sciences

International Commission:

President: B.P. Mansourian, *WHO (ret.) Switzerland*

Vice President: A. Wojtczak, *Institute for International Medical Education, NY, USA*

B. McA. Sayers, *Imperial College of Science, Technology and Medicine, London UK*

S.M. Mahfouz, *Cairo University, Egypt* Members

A. Arata, *USAID, Alexandria, Virginia, USA*

B P.R. Aluwihare, *University of Peradeniya, Kandy, Sri Lanka*

G. W. Brauer, *University of Victoria, Victoria, BC, Canada*

M. R.G. Manciaux, *Nancy University, Vandoeuvre-les-Nancy, France*

J. Szczerban, *Medical University of Warsaw, Warsaw, Poland*

Y. L. G. Verhasselt, *Vrije Universiteit Brussels, Brussels, Belgium*

Alberto Pellegrini, *Ipanema, Rio de Janerio, RJ, Brasil*

Assen Jablensky, *The University of Western Australia, Perth, Australia*

R. Kitney, *Imperial College of Science, Technology and Medicine London, UK*

Arminee Kazanjian, *University of British Columbia, Canada*

Tomris Turmen, *WHO, Turkey*

R. Leidl, *Ludwig Maximilians-University, Munich, Germany*

K. Sorour, *Cairo University, Egypt*
eISBN: 978-1-84826-283-6, 978-1-84826-284-3
ISBN: 978-1-84826-733-6, 978-1-84826-734-3

Biological Science Fundamentals and Systematics (Vols. 1-4)

Editors: Alessandro Minelli, *University of Padova, Italy*

Giancarlo Contrafatto, *University of Natal, South Africa*
eISBN: 978-1-84826-304-8, 978-1-84826-305-5, 978-1-84826-306-2, 978-1-84826-189-1
ISBN: 978-1-84826-754-1, 978-1-84826-755-8, 978-1-84826-756-5, 978-1-84826-639-1

Extremophiles (Vols. 1-3) (Life under extreme environmental conditions)

Editors: Charles Gerday , *University of Liege, Belgium,*

Nicolas Glansdorff, *Vrjie Universiteit Brussel, Belgium*
eISBN: 978-1-905839-93-3, 978-1-905839-94-0, 978-1-905839-95-7
ISBN: 978-1-84826-993-4, 978-1-84826-994-1, 978-1-84826-995-8

Ethnopharmacology (Vols. 1-2)

Editors: Elaine Elisabetsky, *Universidade Federal do Rio Grande do Sul, Brazil*

Nina L. Etkin, *University of Hawai'I, USA*
eISBN: 978-1-905839-96-4, 978-1-905839-97-1
ISBN: 978-1-84826-996-5, 978-1-84826-997-2

Pharmacology

Editor: Harry Majewski, *RMIT University, Victoria, Australia*
eISBN: 978-1-84826-180-8, 978-1-84826-254-6
ISBN: 978-1-84826-630-8, 978-1-84826-704-6

**Biophysics

Editor: Mohamed I.El Gohary,*Al Azhar University, Cairo, Egypt*
eISBN:
ISBN:

**Phytochemistry and Pharmacognosy

Editors: John M. Pezzuto, *University of Hawaii, Hilo, 96720, Hawaii*

Massuo Jorge Kato, *University of Sao Paulo, Sao-Paulo, 05508-000, Brasil*
eISBN:
ISBN:

**Neuroscience

Editor: Jose Masdeu, *National Institutes of Health (CBDB-NIMH), Bethesda, MD*
eISBN:
ISBN:

**History of Medicine

International Commission:

President: Athanasios Diamandopoulos, *Renal department, St. Andrew's regional Hospital, Patras, Greece.*

Vice Presidents:

Carlos Viesca, *UNAM Mexico.*

Alain Lellouch, *DIM de l'Hopital, Saint-Germain-en-Laye, France*

Members:

S. Kotteck, *Israel*

C. Bergdold, *Cologne, Germany*
F. Alecbari, *Atjerbaizan*
B. Fontini, *South America*
T. Sorokina, *Moscow, Russia*
J. Pearns, *Australia*
eISBN:
  ISBN:

**Reproduction and Development Biology

Editor: Andre Pires da Silva, *Department of Biology, The University of Texas at Arlington, USA*
eISBN:
  ISBN:

BIOTECHNOLOGY

Editors: Horst W. Doelle, *MIRCEN-Biotechnology, Australia*

Edgar J. DaSilva, *Section of Life Sciences, Division of Basic and Engineering Sciences, UNESCO, France*

(*The bulleted items in the following represent Topic titles. The total number of chapters is 240. The size of an entry (Chapter) may vary from about 5000 words to about 30000 words.*)

Volume 1: Fundamentals in Biotechnology
eISBN 978-1-84826-255-3
ISBN 978-1-84826-705-3

Volume 2: Methods in Biotechnology
eISBN 978-1-84826-256-0
ISBN 978-1-84826-706-0

Volume 3: Methods in Gene Engineering
eISBN 978-1-84826-257-7
ISBN 978-1-84826-707-7

Volume 4: Bioprocess Engineering - Bioprocess Analysis
eISBN 978-1-84826-258-4
ISDN 978-1-84826 708 4

Volume 5: Industrial Biotechnology - Part I
eISBN 978-1-84826-259-1
ISBN 978-1-84826-709-1

Volume 6: Industrial Biotechnology - Part II
eISBN 978-1-84826-260-7
ISBN 978-1-84826-710-7

Volume 7: Special Processes for Products, Fuel and Energy
eISBN 978-1-84826-261-4
ISBN 978-1-84826-711-4

Volume 8: Agricultural Biotechnology
eISBN 978-1-84826-262-1
ISBN 978-1-84826-712-1

Volume 9: Marine Biotechnology
eISBN 978-1-84826-263-8
ISBN 978-1-84826-713-8

Volume 10: Environmental Biotechnology - Socio-Economic Strategies for Sustainability
eISBN 978-1-84826-264-5
ISBN 978-1-84826-714-5

Volume 11: Medical Biotechnology - Fundamentals and Modern Development – Part I
eISBN 978-1-84826-265-2
ISBN 978-1-84826-715-2

Volume 12: Medical Biotechnology - Fundamentals and Modern Development – Part II
eISBN 978-1-84826-266-9
ISBN 978-1-84826-716-9

Volume 13: Social, Educational and Political Aspects of Biotechnology - An Overview and an Appraisal of Biotechnology in a Changing World – Part I
eISBN 978-1-84826-267-6
ISBN 978-1-84826-717-6

Volume 14: Social, Educational and Political Aspects of Biotechnology - An Overview and an Appraisal of Biotechnology in a Changing World – Part II
eISBN 978-1-84826-268-3
ISBN 978-1-84826-718-3
Volume 15: The Role of microbial resources centers and UNESCO in the Development of Biotechnology
eISBN 978-1-84826-269-0
ISBN 978-1-84826-719-0

TROPICAL BIOLOGY AND CONSERVATION MANAGEMENT

International Commission on Tropical Biology and Conservation Management
President:
Kleber Del Claro, *Universidade Federal de Uberlândia, Uberlândia, MG, Brazil*
Vice Presidents:
Paulo S. Oliveira, *Universidade Estadual de Campinas, Brazil*
Victor Rico-Gray, *Instituto de Ecologia, A.C., México*
Commissioners:
Ana Angélica Almeida Barbosa, *Universidade Federal de Uberlandia, Brazil*
Arturo Bonet, *Instituto de Ecología, A.C., Apartado, Veracruz, México.*
Fábio Rúbio Scarano, *Universidade Federal do Rio de Janeiro, Instituto de Biologia, Brazil*
Francisco José Morales Garzón, *International Center for Tropical Agriculture (CIAT),Cali, Colombia*
Marcus Vinicius Sampaio, *Universidade Federal de Uberlandia, Brazil*
Molly R. Morris, *Ohio University, USA*
Nelson Ramírez, *Universidad Central de Venezuela, Venezuela*
Oswaldo Marcal Júnior, *Universidade Federal de Uberlandia, Brazil*
Regina Helena Ferraz Macedo, *Universidade de Brasilia, Brazil*
Robert J. Marquis, *University of Missouri, USA*
Lisias Coelho- *Federal University of Uberlândia – Agronomy Institute Brazil*
Rogério Parentoni Martins, *Universidade Federal de Minas Gerais, Brazil*
Silvio Carlos Rodrigues, *Universidade Federal de Uberlandia, Brazil*
Ulrich Lüttge, *University of: Techniche Univeritat Darmstadt, Denmark*
(*The bulleted items in the following represent Topic titles. The total number of chapters is 240. The size of an entry (Chapter) may vary from about 5000 words to about 30000 words.*)

Volume 1: Natural History of Tropical Plants
eISBN 978-1-84826-272-0
ISBN 978-1-84826-722-0
Volume 2: Human Impact on Tropical Ecosystems
eISBN 978-1-84826-273-7
ISBN 978-1-84826-723-7
Volume 3: Agriculture
eISBN 978-1-84826-274-4
ISBN 978-1-84826-724-4
Volume 4: Botany
eISBN 978-1-84826-275-1
ISBN 978-1-84826-725-1
Volume 5: Ecology
eISBN 978-1-84826-276-8
ISBN 978-1-84826-726-8
Volume 6: Phytopathology and Entomology-I
eISBN 978-1-84826-277-5
ISBN 978-1-84826-727-5
Volume 7: Phytopathology and Entomology-II
eISBN 978-1-84826-278-2
ISBN 978-1-84826-728-2

Volume 8: Zoology
eISBN 978-1-84826-279-9
ISBN 978-1-84826-729-9

Volume 9: Desert Ecosystems
eISBN 978-1-84826-280-5
ISBN 978-1-84826-730-5

Volume 10: Savannah Ecosystems
eISBN 978-1-84826-281-2
ISBN 978-1-84826-731-2

Volume 11: Case Studies
eISBN 978-1-84826-282-9
ISBN 978-1-84826-732-9

SOCIAL SCIENCES AND HUMANITIES

Peace, Literature, and Art (Vols. 1-2)

Editor: Ada Aharoni, *Technion - Israel Institute of Technology, and Communications Commission of IPRA, Israel*
eISBN: 978-1-84826-076-4, 978-1-84826-077-1
ISBN: 978-1-84826-526-4, 978-1-84826-527-1

**Global Security

Editors: Pinar Bilgin, *Fellow, Woodrow Wilson International Center for Scholars, Washington, DC USA*
Paul D. Williams, *George Washington University, USA*
eISBN:
ISBN:

Environmental Laws and Their Enforcement (Vols. 1-2)

Editors: A. Dan Tarlock, *Illinois Institute of Technology, USA*
eISBN: 978-1-84826-113-6, 978-1-84826-293-5
ISBN: 978-1-84826-563-9, 978-1-84826-743-5

Linguistic Anthropology

Editor: Anita Sujoldzic, *University of Zagreb, Zagreb, Croatia*
eISBN: 978-1-84826-225-6
ISBN: 978-1-84826-675-9

Archaeology (Vols. 1-2)

Editor: Donald L. Hardesty, *University of Nevada, Reno, USA*
eISBN: 978-1-84826-002-3, 978-1-84826-003-0
ISBN: 978-1-84826-452-6, 978-1-84826-453-3

Culture, Civilization and Human Society (Vols. 1-2)

Editors: Herbert Arlt, *Research Institute for Austrian and International Literature and Cultural Studies (INST), Austria,*
Donald G. Daviau, *University of California, Riverside CA, USA*
eISBN: 978-1-84826-190-7, 978-1-84826-191-4
ISBN: 978-1-84826-640-7, 978-1-84826-641-4

Literature and the Fine Arts

Editors: Herbert Arlt, *Research Institute for Austrian and International Literature and Cultural Studies (INST), Austria,*
Donald G. Daviau, *University of California, Riverside CA, USA*
Peter Horn, *University of Cape Town, Cape Town, South Africa*

eISBN: 978-1-84826-122-8
ISBN: 978-1-84826-572-1

**Philosophy and World Problems
Editor: John McMurtry, *University of Guelph, Canada*
eISBN:
ISBN:

Fundamental Economics (Vols. 1-2)
Editor: Mukul Majumdar, *Cornell University, USA*

Ian Wills, *Monash University, Australia*
Pasquale Michael Sgro, *Deakin University, Australia*
John M. Gowdy, *Rensselaer Polytechnic Institute, USA*
eISBN: 978-1-84826-315-4, 978-1-84826-765-7
ISBN:978-1-84826-316-1, 978-1-84826-766-4

Law
Editors: Aaron Schwabach, *Thomas Jefferson School of Law, USA,*
Arthur J. Cockfield *Queen's University, Canada*
eISBN: 978-1-905839-68-1
ISBN: 978-1-848269-68-2

Government and Politics (Vols. 1-2)
Editor: Masashi Sekiguchi, *Tokyo Metropolitan University, Japan*
eISBN: 978-1-905839-69-8, 978-1-905839-70-4
ISBN: 978-1-84826-969-9, 978-1-84826-970-5

Journalism and Mass Communication (Vols. 1-2)
Editor: Rashmi Luthra, *University of Michigan, USA*
eISBN: 978-1-905839-71-1, 978-1-905839-72-8
ISBN: 978-1-84826-971-2, 978-1-84826-972-9

Unity of Knowledge (in Transdisciplinary Research for Sustainability) (Vols. 1-2)
Editor: Gertrude Hirsch Hadorn, *Federal Institute of Technology, ETH Zurich Zentrum, Switzerland*
eISBN: 978-1-905839-82-7, 978-1-905839-83-4
ISBN: 978-1-84826-982-8, 978-1-84826-983-5

Comparative Literature: Sharing Knowledges for Preserving Cultural Diversity
International Commission:
President: Lisa Block de Behar, *University of Republica, Montevideo, Uruguay* (Former President: Tania Franco Carvalhal, *Federal University of Rio Grande do Sul, Porto Alegre, Brazil* who has passed away in 2006)
Vice Presidents: Paola Mildonian, *Universita Ca Foscari di Venezia, Venezia, Italy*
Jean-Michel Djian, *University of Paris 8, France*
Djelal Kadir, *Pennsylvania State University, University Park, PA, USA*
Alfons Knauth, *Ruhr University of Bochum, Germany*
Dolores Romero Lopez, *Universidad of Complutense de Madrid, Spain*
Márcio Seligmann Silva, *University of Campinas, Sao Paulo, Brazil*
eISBN: 978-1-84826-175-4, 978-1-84826-322-2
ISBN: 978-1-84826-625-4, 978-1-84826-772-5

History and Philosophy of Science and Technology (Vols. 1-4)
Editors: Pablo Lorenzano, *Universidad Nacional de Quilmes, Argentina*

Eduardo Ortiz, *Imperial College, London,, UK*

Hans-Jorg Rheinberger, *Max Planck Institute fur Wisenschaftsgeschichte, Berlin, 14195, Germany*

Carlos Delfino Galles, *FCEIA-UNR, Pelligrino 250, Rosario 2000, Argentina*
eISBN: 978-1-84826-323-9, 978-1-84826-324-6, 978-1-84826-325-3, 978-1-84826-326-0
ISBN: 978-1-84826-773-2, 978-1-84826-774-9, 978-1-84826-775-6, 978-1-84826-776-3

Linguistics

Editor: Vesna Muhvic-Dimanovski, *University of Zagreb,*

Lelija Socanac, *Department of Language Research of the Croatian Academy of Sciences*
eISBN: 978-1-84826-287-4
ISBN: 978-1-84826-737-4

Religion, Culture, and Sustainable Development (Vols. 1-3)

Editors: Roberto Blancarte Pimentel, *El Colegio de Mexico, México*
eISBN: 978-1-84826-328-4, 978-1-84826-329-1, 978-1-84826-330-7
ISBN: 978-1-84826-778-7, 978-1-84826-779-4, 978-1-84826-780-0

World System History

Editors: George Modelski, *University of Washington, Settle, Washington WA, USA*

Robert A. Denemark, *University of Delaware, USA*
eISBN: 978-1-84826-218-8
ISBN: 978-1-84826-668-1

**World Civilizations

Editor: Robert Holton, *Trinity College, University of Dublin, Dublin, Ireland*
eISBN:
ISBN:

Historical Developments and Theoretical Approaches in Sociology (Vols. 1-2)

Editor: Charles Crothers, *Auckland University of Technology, New Zealand*
eISBN: 978-1-84826-331-4, 978-1-84826-332-1
ISBN: 978-1-84826-781-7, 978-1-84826-782-4

Nonviolent Alternatives for Social Change

Editor: Ralph V. Summy, *The University of Queensland, Australia*
eISBN: 978-1-84826-220-1
ISBN: 978-1-84826-670-4

Demography (Vols. 1-2)

Editor: Zeng Yi, *Duke University, Durham, USA; Peking University, China*
eISBN: 978-1-84826-307-9, 978-1-84826-308-6
ISBN: 978-1-84826-757-2, 978-1-84826-758-9

PHYSICAL SCIENCES, ENGINEERING AND TECHNOLOGY RESOURCES

Development of Physics

Editor: Gyo Takeda, *University of Tokyo and Tohoku University, Japan*
eISBN: 978-1-84826-201-0
ISBN: 978-1-84826-651-3

Fundamentals of Physics (Vols. 1-3)

Editor: José Luis Morán López, *Instituto Potosino de Investigación Científica y Tecnológica, Mexico*
eISBN: 978-1-905839-52-0, 978-1-905839-53-7, 978-1-84826-223-2

ISBN: 978-1-84826-952-1, 978-1-84826-953-8, 978-1-84826-673-5

Physical Methods, Instruments and Measurements (Vols. 1-4)

Editor: Yuri Mikhailovitsh Tsipenyuk, *P.L. Kapitza Institute for Physical Problems (RAS), Russia*
eISBN: 978-1-905839-54-4, 978-1-905839-55-1, 978-1-905839-56-8, 978-1-905839-57-5
ISBN: 978-1-84826-954-5, 978-1-84826-955-2, 978-1-84826-956-9, 978-1-84826-957-6

Mechanical Engineering, Energy Systems and Sustainable Development (Vols. 1-5)

Editors: Konstantin V. Frolov, *Mechanical Engineering Research Institute, Russian Academy of Sciences, Russia*

R.A. Chaplin, *University of New Brunswick, Canada*

Christos Frangopoulos, *National Technical University of Athens, Greece*
eISBN: 978-1-84826-294-2, 978-1-84826-295-9, 978-1-84826-296-6, 978-1-84826-297-3, 978-1-84826-298-0
ISBN: 978-1-84826-744-2, 978-1-84826-745-9, 978-1-84826-746-6, 978-1-84826-747-3, 978-1-84826-748-0

Materials Science and Engineering (Vols. 1-3)

Editor: Rees D. Rawlings, *Imperial College of Science, Technology and Medicine, UK*
eISBN: 978-1-84826-032-0, 978-1-84826-033-7, 978-1-84826-034-4
ISBN: 978-1-84826-482-3, 978-1-84826-483-0, 978-1-84826-484-7

Civil Engineering (Vols. 1-2)

Editors: Kiyoshi Horikawa, *Musashi Institute of Technology, Japan*

Qizhong Guo, *Rutgers, The State University of New Jersey, USA*
eISBN: 978-1-905839-73-5, 978-1-905839-74-2
ISBN: 978-1-84826-973-6, 978-1-84826-974-3

Electrical Engineering (Vols. 1-3)

Editor: Kit Po Wong, *Hong Kong Polytechnic University, Hong Kong*
eISBN: 978-1-905839-77-3, 978-1-905839-78-0, 978-1-905839-79-7
ISBN: 978-1-84826-977-4, 978-1-84826-978-1, 978-1-84826-979-8

Transportation Engineering and Planning (Vols. 1-2)

Editor: Tschangho John Kim, *University of Illinois at Urbana-Champaign, USA*
eISBN: 978-1-905839-80-3, 978-1-905839-81-0
ISBN: 978-1-84826-980-4, 978-1-84826-981-1

Telecommunication Systems and Technologies (Vols. 1-2)

Editor: Paolo Bellavista, *DEIS-LIA, Universita degli Studi di Bologna, Italy*
eISBN: 978-1-84826-000-9, 978-1-84826-001-6
ISBN: 978-1-84826-450-2, 978-1-84826-451-9

**Structural Engineering and Structural Mechanics

Editor: Xila S. Liu, *Tsinghua University, Beijing, CHINA*
eISBN:
ISBN:

**Structural Engineering and Geomechanics

Editor: Sashi K. Kunnath, *University of California, DAVIS, USA*
eISBN:
ISBN:

**Crystallography

Editors: John R. Helliwell, *The University of Manchester, Manchester, UK*

Santiago Gracia Granda, *University of Ovieddo, Spain*
eISBN:
ISBN:

**Nanoscience and Nanotechnologies

Editors: Valeri Nikolayevich Harkin, *International Commission, Moscow, 115419, Russia*
Chunli Bai, *Institute of Chemistry, Chinese Academy of Sciences, Beijing, 100864, China*
Sae-Chul Kim, *International Union of Pure and Applied Chemistry, South Korea*
eISBN:
ISBN:

**Continuum Mechanics

Editors: José Merodio, *Universidad Politécnica de Madrid, Madrid, Spain*
Giuseppe Saccomandi, *Sezione di Ingegneria Industriale, Dipartimento di Ingegneria dell'Innovazione, niversita degli Studi di Lecce, Via per Monteroni, 73100 Lecce, Italy*
eISBN:
ISBN:

**Biomechanics

Editors: Manuel Doblare, *Universidad de Zaragoza, Spain*
Jose Merodio,*Universidad Politécnica de Madrid, Madrid, Spain*
José Ma Goicolea Ruigómez, *Ciudad Universitaria, Madrid, Spain*
eISBN:
ISBN:

**Building Services Engineering

Editor: Yuguo Li, *The University of Hong Kong, Hong Kong, China.*
eISBN:
ISBN:

**Pressure Vessels and Piping Systems

Editors: Yong W.Kwon, *Naval Postgraduate School, Monterey, CA 93943 USA*
Poh-Sang Lam, *Savannah River National Laboratory, Aiken, South Carolina 29808, USA*
eISBN:
ISBN:

**Tribology: Friction, Wear, Lubrication

Editor: Roberto Bassani, *Universita di Pisa, Pisa, Italy*
eISBN:
ISBN:

**Welding Engineering and Technology

Editors: Slobodan Kralj, *Faculty of Mechanical Engineering and Naval Architecture, Zagreb, Croatia*
Branko Bauer, *Dept. of Welded Structures, Zagreb, Zagreb, 10000, Croatia*
eISBN:
ISBN:

**Ships and Offshore Structures

Editors: Jeom Kee Paik, *Pusan National University , Geumjeong-Gu, Busan 609-735, Korea*
Ge(George) Wang, *American Bureau of Shipping Technology,,, USA*
International Editorial Advisory Board (IEAB): Weicheng Cui, *China;* Paul A. Frieze, *UK;* Yoshiaki Kodama, *Japan;* David Molyneux, *Canada;* Anil K. Thayamballi, *USA;* P.A. Wilson, *UK*

eISBN:
ISBN:

**Cold Regions Science and Marine Technology
Editor: Hayley Shen, *Clarkson University, Potsdam, USA*
eISBN:
ISBN:

**Experimental Mechanics
Editor: José Luiz de França Freire, *Endereço profissional Pontifícia Universidade Católica do Rio de Janeiro, Brasil.*
eISBN:
ISBN:

CONTROL SYSTEMS, ROBOTICS, AND AUTOMATION

(Vols. 1-22)

Editor: Heinz Unbehauen, *Ruhr-Universität Bochum, Germany*

(The bulleted items in the following represent Topic titles. The total number of chapters is 240. The size of an entry (Chapter) may vary from about 5000 words to about 30000 words.)

Volume 1: System Analysis and Control: Classical Approaches I
- *Control Systems, Robotics and Automation-Introduction and Overview*
- *Elements of Control Systems*
- *Stability Concepts*

eISBN 978-1-84826-140-2
ISBN 978-1-84826-590-5

Volume 2: System Analysis and Control: Classical Approaches II
- *Classical Design Methods for Continuous LTI-Systems*
- *Digital Control Systems*

eISBN 978-1-84826-141-9
ISBN 978-1-84826-591-2

Volume 3: System Analysis and Control: Classical Approaches III
- *Design of State Space Controllers (Pole Placement) for SISO Systems*
- *Basic Nonlinear Control Systems*

eISBN 978-1-84826-142-6
ISBN 978-1-84826-592-9

Volume 4: Modeling and System Identification I
- *Modeling and Simulation of Dynamic Systems*

eISBN 978-1-84826-143-3
ISBN 978-1-84826-593-6

Volume 5: Modeling and System Identification II
- *Frequency Domain System Identification*
- *Identification of Linear Systems in Time Domain*

eISBN 978-1-84826-144-0
ISBN 978-1-84826-594-3

Volume 6: Modeling and System Identification III
- *Identification of Nonlinear Systems*
- *Bound-based Identification*
- *Practical Issues of System Identification*

eISBN 978-1-84826-145-7
ISBN 978-1-84826-595-0

Volume 7: Advanced Control Systems I

- *Supervisory Distributed Computer Control Systems*
- *Automation and Control of Thermal Processes*
- *Automation and Control of Electrical Power Generation and Transmission Systems*
  eISBN 978-1-84826-157-0
  ISBN 978-1-84826-607-0

Volume 19: Industrial Applications of Control Systems II
- *Automation and Control in Process Industries*
- *Automation and Control in Production Processes*
  eISBN 978-1-84826-158-7
  ISBN 978-1-84826-608-7

Volume 20: Industrial Applications of Control Systems III
- *Automation and Control in Traffic Systems*
  eISBN 978-1-84826-159-4
  ISBN 978-1-84826-609-4

Volume 21: Elements of Automation and Control
  eISBN 978-1-84826-160-0
  ISBN 978-1-84826-610-0

Volume 22: Robotics
  eISBN 978-1-84826-161-7
  ISBN 978-1-84826-611-7

CHEMICAL SCIENCES, ENGINEERING AND TECHNOLOGY RESOURCES

Fundamentals of Chemistry (Vols. 1-2)

Editor: Sergio Carra, *Politecnico di Milano, Italy*
  eISBN: 978-1-905839-58-2, 978-1-905839-59-9
  ISBN: 978-1-84826-958-3, 978-1-84826-959-0

Environmental and Ecological Chemistry (Vols. 1-3)

Editor: Aleksandar Sabljic, *Rudjer Boskovic Institute, Croatia*
  eISBN: 978-1-84826-186-0, 978-1-84826-206-5, 978-1-84826-212-6
  ISBN: 978-1-84826-692-6, 978-1-84826-693-3, 978-1-84826-694-0

**Chemical Engineering and Chemical Process Technology

Editors: Ryszhard Pohorecki, *Warsaw University of Technology,*
  John Bridgwater, *University of Cambridge,*
  Rafiqul GANI, *Technical University of Denmark*
  eISBN:
  ISBN:

Chemical Ecology

Editor: Jorg D. Hardege, *University of Hull, UK*
  eISBN: 978-1-84826-179-2
  ISBN: 978-1-84826-629-2

Inorganic and Bio-inorganic Chemistry (Vols. 1-2)

Editor: Ivano Bertini, *University of Florence, Florence, Italy*
  eISBN: 978-1-84826-214-0, 978-1-84826-215-7
  ISBN: 978-1-84826-664-3, 978-1-84826-665-0

Organic and Bio-molecular Chemistry (Vols. 1-2)

Editor: Francesco Nicotra, *University of Milano Bicocca, Italy*
  eISBN: 978-1-905839-98-8, 978-1-905839-99-5

ISBN: 978-1-84826-998-9, 978-1-84826-999-6

Radiochemistry and Nuclear Chemistry (Vols. 1-2)

Editor: Sándor Nagy, *Eotvos Lorand University, Hungary*
eISBN: 978-1-84826-126-6, 978-1-84826-127-3
ISBN: 978-1-84826-576-9, 978-1-84826-577-6

**Electrochemistry

Editors: Juan M. Feliu-Martinez, *Universidad de Alicante, Departmento de Química Física, Spain*
Victor Climent Payá, *Universidad de Alicante, Departmento de Química Física, Spain*
eISBN:
ISBN:

**Catalysis

Editor: Gabriele Centi, *Universita di Messina, Salita Sperone, Messina, Italy.*
eISBN:
ISBN:

Rheology (Vols. 1-2)

Editor: Crispulo Gallegos, *Universidad de Huelva, Spain*
eISBN: 978-1-84826-319-2, 978-1-84826-320-8
ISBN:978-1-84826-765-9, 978-1-84826-770-1

**Phytochemistry and Pharmacognosy

Editors: John M. Pezzuto, *University of Hawaii at Hilo, Hawaii, USA*
Massuo Jorge Kato, *Instituto de Química USP, Av. Prof. Lineu Prestes, São Paulo, Brasil*
eISBN:
ISBN:

WATER SCIENCES, ENGINEERING AND TECHNOLOGY RESOURCES

The Hydrological Cycle (Vols. 1-4)

Editor: Igor Alekseevich Shiklomanov, *State Hydrological Institute (SHI), Russia*
eISBN: 978-1-84826-024-5, 978-1-84826-025-2, 978-1-84826-026-9, 978-1-84826-193-8
ISBN: 978-1-84826-474-8, 978-1-84826-475-5, 978-1-84826-476-2, 978-1-84826-643-8

Types and Properties of Water (Vols. 1-2)

Editor: Martin Gaikovich Khublaryan, *Water Problems Institute, Russian Academy of Science, Russia*
eISBN: 978-1-90583-922-3, 978-1-90583-923-0
ISBN: 978-1-84826-922-4, 978-1-84826-923-1

Fresh Surface Water (Vols. 1-3)

Editor: James C.I. Dooge, *University College Dublin, Ireland*
eISBN: 978-1-84826-011-5, 978-1-84826-012-2, 978-1-84826-013-9
ISBN: 978-1-84826-461-8, 978-1-84826-462-5, 978-1-84826-463-2

Groundwater (Vols. 1-3)

Editors: Luis Silveira, Eduardo J. Usunoff, *Uruguay Instituto de Hidrologia de Llanuras, Argentina*
eISBN: 978-1-84826-027-6, 978-1-84826-028-3, 978-1-84826-029-0
ISBN: 978-1-84826-477-9, 978-1-84826-478-6, 978-1-84826-479-3

Water Storage, Transport, and Distribution (Vols. 1-2)

Editor: Yutaka Takahasi, *University of Tokyo, Japan*
eISBN: 978-1-84826-176-1

ISBN: 978-1-84826-626-1

Wastewater Recycling, Reuse, and Reclamation (Vols. 1-2)

Editor: Saravanamuthu (Vigi) Vigneswaran, *University of Technology, Sydney, Australia*

eISBN: 978-1-90583-924-7, 978-1-90583-925-4
ISBN: 978-1-84826-924-8, 978-1-84826-925-5

Hydraulic Structures, Equipment and Water Data Acquisition Systems (Vols. 1-4)

Editors: Jan Malan Jordaan, *University of Pretoria, South Africa*

Alexander Bell, *Bell & Associates, South Africa*

eISBN: 978-1-84826-049-8, 978-1-84826-050-4, 978-1-84826-051-1, 978-1-84826-192-1
ISBN: 978-1-84826-499-1, 978-1-84826-500-4, 978-1-84826-501-1, 978-1-84826-642-1

Water Resources Management (Vols. 1-2)

Editors: Hubert H.G. Savenije and Arjen Y. Hoekstra, *International Institute for Infrastructural, Hydraulic and Environmental Engineering (IHE - Delft), The Netherlands*

eISBN: 978-1-84826-177-8, 978-1-84826-224-9
ISBN: 978-1-84826-627-8, 978-1-84826-674-2

Water Quality and Standards (Vols. 1-2)

Editors: Shoji Kubota, *Kyushyu University, Japan*

Yoshiteru Tsuchiya, *Kogakuin University, Tokyo, Japan*

eISBN: 978-1-84826-030-6, 978-1-84826-031-3
ISBN: 978-1-84826-480-9, 978-1-84826-481-6

Environmental and Health Aspects of Water Treatment and Supply

Editors: Shoji Kubota, *Kyushyu University, Japan*

Yasumoto Magara, *Hokkaido University, Japan*

eISBN: 978-1-84826-178-5
ISBN: 978-1-84826-628-5

Water-Related Education, Training and Technology Transfer

Editor: Andre van der Beken, *Vrije Universitiet Brussel (VUB), Belgium*

eISBN: 978-1-84826-015-3
ISBN: 978-1-84826-465-6

Water Interactions with Energy, Environment and Food & Agriculture (Vols. 1-2)

Editor: Maria Concepcion Donoso, *Water Center for the Humid Tropics of Latin America and the Caribbean (CATHALAC), Panama*

eISBN: 978-1-84826-172-3, 978-1-84826-196-9
ISBN: 978-1-84826-622-3, 978-1-84826-646-9

Water and Development (Vols. 1-2)

Editor: Catherine M. Marquette, *Christian Michelsen Institute, Development Studies and Human Rights, Norway*

eISBN: 978-1-84826-173-0, 978-1-84826-197-6
ISBN: 978-1-84826-623-0, 978-1-84826-647-6

Future Challenges of Providing High-Quality Water (Vols. 1-2)

Editors: Jo-Ansie van Wyk, *University of South Africa (UNISA), South Africa*

Richard Meissner, *University of Pretoria, South Africa*

Hannatjie Jacobs, *Independent Research Consultant, South Africa*

eISBN: 978-1-84826-209-6, 978-1-84826-230-0
ISBN: 978-1-84826-659-9, 978-1-84826-680-3

Hydrological Systems Modeling (Vols. 1-2)

Editors: Lev S. Kuchment, *Water Problems Institute, Russian Academy of Sciences, Moscow, Russia*

Vijay P. Singh, *Texas A & M University*
eISBN: 978-1-84826-198-3, 978-1-84826-187-7
ISBN: 978-1-84826-648-3, 978-1-84826-637-7

Wastewater Treatment Technologies (Vols. 1-3)

Editor: Saravanamuthu (Vigi) Vigneswaran, *University of Technology, Sydney, Australia*
eISBN: 978-1-84826-188-4, 978-1-84826-194-5, 978-1-84826-250-8
ISBN: 978-1-84826-638-4, 978-1-84826-644-5, 978-1-84826-700-8

ENERGY SCIENCES, ENGINEERING AND TECHNOLOGY RESOURCES

Coal, Oil Shale, Natural Bitumen, Heavy Oil and Peat (Vols. 1-2)

Editor: Gao Jinsheng, *East China University of Science and Technology(ECUST), China*
eISBN: 978-1-84826-017-7, 978-1-84826-018-4
ISBN: 978-1-84826-467-0, 978-1-84826-468-7

Nuclear Energy Materials and Reactors (Vols. 1-2)

Editors: Yassin A. Hassan, *Texas A&M University, USA*

Robin A. Chaplin, *University of New Brunswick, Canada*
eISBN: 978-1-84826-311-6, 978-1-84826-761-9
ISBN:978-1-84826-312-3, 978-1-84826-762-6

Renewable Energy Sources Charged with Energy from the Sun and Originated from Earth-Moon Interaction (Vols. 1-2)

Editor: Evald Emilievich Shpilrain, *IVTAN Institute for High Temperatures, Russian Academy of Sciences, Russia*
eISBN: 978-1-84826-019-1, 978-1-84826-020-7
ISBN: 978-1-84826-469-4, 978-1-84826-470-0

Thermal Power Plants (Vols. 1-3)

Editor: Robin A. Chaplin, *University of New Brunswick, Canada*
eISBN: 978-1-905839-26-1, 978-1-905839-27-8, 978-1-905839-28-5
ISBN: 978-1-84826-926-2, 978-1-84826-927-9, 978-1-84826-928-6

Thermal to Mechanical Energy Conversion(Vols. 1-3)

Editor: Oleg N. Favorsky, *Department of Physical and Technical Problems of Energetics, Russian Academy of Sciences, Russia*
eISBN: 978-1-84826-021-4, 978-1-84826-022-1, 978-1-84826-023-8
ISBN: 978-1-84826-471-7, 978-1-84826-472-4, 978-1-84826-473-1

Energy Carriers and Conversion Systems with Emphasis on Hydrogen (Vols. 1-2)

Editor: Tokio Ohta, *Yokohama National University and Chairman of Frontier Information and Learning Organization (FILO), Japan*
eISBN: 978-1-905839-29-2, 978-1-905839-30-8
ISBN: 978-1-84826-929-3, 978-1-84826-930-9

Energy Storage Systems (Vols. 1-2)

Editor: Yalsin Gogus, *Middle East Technical University Ankara, Turkey*
eISBN: 978-1-84826-162-4, 978-1-84826-163-1
ISBN: 978-1-84826-612-4, 978-1-84826-613-1

Air Conditioning - Energy Consumption and Environmental Quality

Editor: Matheos Santamouris, *University of Athens, Greece*
eISBN: 978-1-84826-169-3
ISBN: 978-1-84826-619-3

Efficient Use and Conservation of Energy (Vols. 1-2)

Editor: Clark W. Gellings, *Technology Initiatives Electric Power Research Institute (EPRI), USA*
eISBN: 978-1-905839-31-5, 978-1-905839-32-2
ISBN: 978-1-84826-931-6, 978-1-84826-932-3

Exergy, Energy System Analysis, and Optimization (Vols. 1-3)

Editor: Christos A. Frangopoulos, *National Technical University of Athens, Greece*
eISBN: 978-1-84826-164-8, 978-1-84826-165-5, 978-1-84826-166-2
ISBN: 978-1-84826-614-8, 978-1-84826-615-5, 978-1-84826-616-2

Energy Policy

Editor: Anthony David Owen, *The University of New South Wales, Australia*
eISBN: 978-1-905839-33-9
ISBN: 978-1-84826-933-0

Solar Energy Conversion and Photoenergy Systems (Vols. 1-2)

Editors: Julian Blanco Gálvez and Sixto Malato Rodríguez, *Plataforma Solar de Almería, Spain*
eISBN: 978-1-84826-285-0, 978-1-84826-286-7
ISBN: 978-1-84826-735-0, 978-1-84826-736-7

**Heat and Mass Transfer

Editors: Tao Wen-Quan, *Xian Jiaotong University, Xian, 710049, PR China*
Qianjin J. Yue, *Dalian U.Technology, China*
eISBN:
ISBN:

**Petroleum Engineering – Downstream

Editor: Pedro de Alcantara Pessoa Filho, *University of Sao* Paulo, *Brazil*
eISBN:
ISBN:

**Petroleum: Refining, Fuels and Petrochemicals

Editor: James G. Speight, *Western Research Institute, Laramie, WY, USA*
eISBN:
ISBN:

**Pipeline Engineering

Editor: Yufeng F. Cheng, *University of Calgary, Calgary, Alberta, Canada*
eISBN:
ISBN:

**Petroleum Engineering – Upstream

Editors: Ezio Mesini and Paolo Macini, *University of Bologna, Italy*
eISBN:
ISBN:

**Energy and Fuel Sciences

Editor: Semih Eser, *Pennsylvania State University, Pennsylvania, USA*
ISBN:

ENVIRONMENTAL AND ECOLOGICAL SCIENCES, ENGINEERING AND TECHNOLOGY RESOURCES

Hazardous Waste

Editors: Domenico Grasso, Timothy M. Vogel and Barth Smets, *Smith College, USA*
eISBN: 978-1-84826-055-9
ISBN: 978-1-84826-505-9

Natural and Human Induced Hazards and Environmental Waste Management (Vols. 1-4)

Editor: Domenico Grasso, *Smith College, USA*
eISBN: 978-1-84826-299-7, 978-1-84826-300-0, 978-1-84826-301-7, 978-1-84826-302-4
ISBN: 978-1-84826-749-7, 978-1-84826-750-3, 978-1-84826-751-0, 978-1-84826-752-7

Coastal Zones and Estuaries

Editors: Federico Ignacio Isla, and Oscar Iribarne, *Universidad Nacional de Mar del Plata, Argentina*
eISBN: 978-1-84826-016-0
ISBN: 978-1-84826-466-3

Marine Ecology

Editors: Carlos M. Duarte, *CSIC-Univ. Illes Balears, Spain*
Antonio Lot, *Universidad Nacional Autonoma de Mexico (UNAM), México*
eISBN: 978-1-84826-014-6
ISBN: 978-1-84826-464-9

Point Sources of Pollution: Local Effects and Their Control (Vols. 1-2)

Editor: Qian Yi, *Tsinghua University, China*
eISBN: 978-1-84826-167-9, 978-1-84826-168-6
ISBN: 978-1-84826-617-9, 978-1-84826-618-6

Environmental Toxicology and Human Health (Vols. 1-2)

Editor: Tetsuo Satoh, *Biomedical Research Institute and Chiba University, Japan*
eISBN: 978-1-84826-251-5, 978-1-84826-252-2
ISBN: 978-1-84826-701-5, 978-1-84826-702-2

Waste Management and Minimization

Editors: Stephen R. Smith, Chris Cheeseman, and Nick Blakey, *Imperial College, UK.*
eISBN: 978-1-84826-119-8
ISBN: 978-1-84826-569-1

Pollution Control Technologies (Vols. 1-3)

Editor: Georgi Stefanov Cholakov, *University of Chemical Technology and Metallurgy, Bulgaria*
Bhaskar Nath, *European Centre for Pollution Research, UK*
eISBN: 978-1-84826-116-7, 978-1-84826-117-4, 978-1-84826-118-1
ISBN: 978-1-84826-566-0, 978-1-84826-567-7, 978-1-84826-568-4

Environmental Systems (Vols. 1-2)

Editor: Achim Sydow, *GMD FIRST, Berlin, Germany*
eISBN: 978-1-84826-210-2, 978-1-84826-211-9
ISBN: 978-1-84826-660-5, 978-1-84826-661-2

Environmental Regulations and Standard Setting

Editor: Bhaskar Nath, *European Centre for Pollution Research, UK*
eISBN: 978-1-84826-103-7
ISBN: 978-1-84826-553-0

Interactions: Energy/Environment

Editor: Jose' Goldemberg, *The University of Sao Paulo, Brazil*
eISBN: 978-1-84826-090-0
ISBN: 978-1-84826-540-0

Biodiversity: Structure and Function (Vols. 1-2)

Editors: Wilhelm Barthlott, *Rheinische Friedrich-Wilhelms-Universitat Bonn, Germany*

K. Eduard Linsenmair, *Universität Würzburg, Germany*

Stefan Porembski, *Universität Rostock, Germany*
eISBN: 978-1-905839-34-6, 978-1-905839-35-3
ISBN: 978-1-84826-934-7, 978-1-84826-935-4

Environmental Monitoring (Vols. 1-2)

Editors: Hilary I. Inyang and John L. Daniels, *The University of North Carolina at Charlotte, USA*
eISBN: 978-1-905839-75-9, 978-1-905839-76-6
ISBN: 978-1-84826-975-0, 978-1-84826-976-7

**Engineering Geology, Environmental Geology & Mineral Economics

Editors: Syed E. Hasan, *University of Missouri, USA*

Syed M. Hasan, *Canada*
eISBN:
ISBN:

Ecology (Vols. 1-2)

Editors: Antonio Bodini, *University of Parma, Italy*

Stefan Klotz, *Centre for Environmental Research Leipzig-Halle Ltd., Germany*
eISBN: 978-1-84826-290-4, 978-1-84826-291-1
ISBN: 978-1-84826-740-4, 978-1-84826-741-1

**Ethnobiology

Editor: John Richard Stepp, *University of Florida, USA*
eISBN:
ISBN:

**Economic Botany

Editor: Brad Bennett, *Florida International University, Miami, Florida, USA*
eISBN:
ISBN:

**Air, Water, Sediment and Soil Pollution Modeling

Editor: Bhaskar Nath, *European Center for Pollution Research, UK*
eISBN:
ISBN:

**Phytotechnologies solutions for sustainable land management

Editor: Tomas Vanek, *Institute of Experimental Botany Academy of Sciences of the Czech Republic and Research Institute of Crop Sciences, Prague Czech Republic.*

eISBN:

ISBN:

**World Environmental History

Editors: Mauro Agnoletti, *University of Florence, Florence Italy; Università degli Studi di Firenze, Italy*

Elizabeth Johann, *International Union of Forest Research Organization, Vienna, 1130, Austria*

Simone Neri Serneri, *University of Siena, 10-53100, Italy*

eISBN:

ISBN:

FOOD AND AGRICULTURAL SCIENCES, ENGINEERING AND TECHNOLOGY RESOURCES

**Soils, Plant Growth and Corp Production

Editor: Willy H. Verheye, *University of Gent, Belgium*

eISBN:

ISBN:

**Interactions: Food and Agriculture/ Environment

Editor: G.Lysenko, *Division of Economics and Landscape Relationships, RAS, Moscow, Russia*

eISBN:

ISBN:

The Role of Food, Agriculture, Forestry and Fisheries in Human Nutrition (Vols. 1-4)

Editor: Victor R. Squires, *Dry Land Management Consultant, Australia*

eISBN: 978-1-84826-134-1, 978-1-84826-135-8, 978-1-84826-136-5, 978-1-84826-195-2

ISBN: 978-1-84826-584-4, 978-1-84826-585-1, 978-1-84826-586-8, 978-1-84826-645-2

Cultivated Plants, Primarily as Food Sources (Vols. 1-2)

Editor: Gyorgy Fuleky, *University of Agricultural Sciences, Hungary*

eISBN: 978-1-84826-100-6, 978-1-84826-101-3

ISBN: 978-1-84826-550-9, 978-1-84826-551-6

Forests and Forest Plants (Vols. 1-3)

Editors: John N. Owens, *Centre for Forest Biology, University of Victoria, Canada*

H. Gyde Lund, *Forest Information Services, USA*

eISBN: 978-1-905839-38-4, 978-1-905839-39-1, 978-1-905839-40-7

ISBN: 978-1-84826-938-5, 978-1-84826-939-2, 978-1-84826-940-8

Fisheries and Aquaculture (Vols. 1-5)

Editor: Patrick Safran, *Centre de Coopération Internationale en Recherche Agronomique pour le Développement (CIRAD), Paris, France, France*

eISBN: 978-1-84826-108-2, 978-1-84826-109-9, 978-1-84826-110-5, 978-1-84826-111-2, 978-1-84826-112-9

ISBN: 978-1-84826-558-5, 978-1-84826-559-2, 978-1-84826-560-8, 978-1-84826-561-5, 978-1-84826-562-2

Food Quality and Standards (Vols. 1-3)

Editor: Radomir Lasztity, *Budapest University of Technology and Economics, Hungary*

eISBN: 978-1-905839-41-4, 978-1-905839-42-1, 978-1-905839-43-8

ISBN: 978-1-84826-941-5, 978-1-84826-942-2, 978-1-84826-943-9

Agricultural Land Improvement: Amelioration and Reclamation (Vols. 1-2)

Editor: Boris Stepanovich Maslov, *Russian Academy of Agriculture Sciences, Russia*
eISBN: 978-1-84826-098-6, 978-1-84826-099-3
ISBN: 978-1-84826-548-6, 978-1-84826-549-3

Food Engineering (Vols. 1-4)

Editor: Gustavo V. Barbosa-Cánovas, *Washington State University, USA*
eISBN: 978-1-905839-44-5, 978-1-905839-45-2, 978-1-905839-46-9, 978-1-905839-47-6
ISBN: 978-1-84826-944-6, 978-1-84826-945-3, 978-1-84826-946-0, 978-1-84826-947-7

Agricultural Mechanization & Automation (Vols. 1-2)

Editors: Paul McNulty and Patrick M. Grace, *University College Dublin, Ireland*
eISBN: 978-1-84826-096-2, 978-1-84826-097-9
ISBN: 978-1-84826-546-2, 978-1-84826-547-9

Management of Agricultural, Forestry and Fisheries Enterprises (Vols. 1-2)

Editor: Robert J. Hudson, *University of Alberta, Canada*
eISBN: 978-1-84826-199-0, 978-1-84826-200-3
ISBN: 978-1-84826-649-0, 978-1-84826-650-6

Animal and Plant Productivity

Editor: Robert J. Hudson, *University of Alberta, Canada*
eISBN: 978-1-84826-317-8
ISBN: 978-1-84826-767-1

Systems Analysis and Modeling in Food and Agriculture

Editors: K. C. Ting, *The Ohio State University, USA*

David H. Fleisher, *United States Department of Agriculture, Alternate Crops and Systems Laboratory, USA*

Luis F. Rodriguez, *National Research Council, NASA Lyndon B. Johnson Space Center, USA*
eISBN: 978-1-84826-133-4
ISBN: 978-1-84826-583-7

Public Policy in Food and Agriculture

Editor: Azzeddine Azzam, *University of Nebraska, Lincoln, USA*
eISBN: 978-1-84826-095-5
ISBN: 978-1-84826-545-5

Impacts of Agriculture on Human Health and Nutrition (Vols. 1-2)

Editors: Ismail Cakmak, *Sabanci University, Turkey*

Ross M. Welch, *Cornell University, USA*
eISBN: 978-1-84826-093-1, 978-1-84826-094-8
ISBN: 978-1-84826-543-1, 978-1-84826-544-8

Interdisciplinary and Sustainability Issues in Food and Agriculture

Editor: Olaf Christen, *Martin-Luther-University, Germany*
eISBN: 978-1-84826-321-5
ISBN: 978-1-84826-771-8

Veterinary Science

Editors: Robert J. Hudson, *University of Alberta, Canada*

Ole Nielson, *University of Guelph, Canada*

J. Bellamy, *University of Prince Edward Island, Canada*

Craig Stephen, *University of Calgary, Canada*

eISBN: 978-1-84826-318-5
ISBN: 978-1-84826-768-8

Agricultural Sciences (Vols. 1-2)

Editor: Rattan Lal, *The Ohio State University, USA*
eISBN: 978-1-84826-091-7, 978-1-84826-092-4
ISBN: 978-1-84826-541-7, 978-1-84826-542-4

**Range and Animal Sciences and Resources Management

Editor: Victor R. Squires, *Dry Land Management Consultant, Australia*
eISBN:
ISBN:

Shaaka Paakam- Vegetarian Culinary Culture of Telugu (Andhra) Draavida Community of South India

Author: Meenakshi Ganti, *PO Box 2623, Abu Dhabi, UAE*
eISBN: 978-1-84826-253-9
ISBN: 978-1-84826-703-9

HUMAN RESOURCES POLICY, DEVELOPMENT AND MANAGEMENT

Human Resources and Their Development (Vols. 1-2)

Editor: Michael J. Marquardt, *The George Washington University, USA*
eISBN: 978-1-84826-056-6, 978-1-84826-057-3
ISBN: 978-1-84826-506-6, 978-1-84826-507-3

Social and Cultural Development of Human Resources

Editor: Tomoko Hamada, *College of William and Mary, USA*
eISBN: 978-1-84826-087-0
ISBN. 978-1-84826-537 0

Quality of Human Resources: Education (Vols. 1-3)

Editors: Natalia Pavlovna Tarasova, *D. Mendeleyev University of Chemical Technology of Russia, Russia*
eISBN: 978-1-905839-07-0, 978-1-905839-08-7, 978-1-905839-09-4
ISBN: 978-1-84826-907-1, 978-1-84826-908-8, 978-1-84826-909-5

Population and Development: Challenges and Opportunities

Editor: Anatoly G. Vishnevsky, *Institute for Economic Forecasting, Russian Academy of Sciences, Russia*
eISBN: 978-1-84826-086-3
ISBN: 978-1-84826-536-3

Quality of Human Resources: Gender and Indigenous Peoples

Editor: Eleonora Barbieri-Masini, *Faculty of Social Sciences, Gregorian University, Italy*
eISBN: 978-1-905839-10-0
ISBN: 978-1-84826-753-4

Environmental Education and Awareness (Vols. 1-2)

Editor: Bhaskar Nath, *European Centre for Pollution Research, UK*
eISBN: 978-1-84826-114-3, 978-1-84826-115-0
ISBN: 978-1-84826-564-6, 978-1-84826-565-3

Sustainable Human Development in the Twenty-First Century (Vols. 1-2)

Editor: Ismail Sirageldin, *The Johns Hopkins University, USA*
eISBN: 978-1-905839-84-1, 978-1-905839-85-8

ISBN: 978-1-84826-984-2, 978-1-84826-985-9

Education for Sustainability

Editors: Robert V. Farrell, *Florida International University (FIU), USA*
eISBN: 978-1-84826-124-2
ISBN: 978-1-84826-574-5

NATURAL RESOURCES POLICY AND MANAGEMENT

Earth System: History and Natural Variability (Vols. 1-4)

Editor: Vaclav Cilek, *Geological Institute, Czech Academy of Sciences, Czech Republic,*
eISBN: 978-1-84826-104-4, 978-1-84826-105-1, 978-1-84826-106-8, 978-1-84826-107-5
ISBN: 978-1-84826-554-7, 978-1-84826-555-4, 978-1-84826-556-1, 978-1-84826-557-8

Climate Change, Human Systems and Policy (Vols. 1-3)

Editor: Antoaneta Yotova, *National Institute of Meteorology and Hydrology, Bulgaria*
eISBN: 978-1-905839-02-5, 978-1-905839-03-2, 978-1-905839-04-9
ISBN: 978-1-84826-902-6, 978-1-84826-903-3, 978-1-84826-904-0

Oceans and Aquatic Ecosystems (Vols. 1-2)

Editor: Eric Wolanski, *FTSE, Australian Institute of Marine Science, Australia*
eISBN: 978-1-905839-05-6, 978-1-905839-06-3
ISBN: 978-1-84826-905-7, 978-1-84826-906-4

Natural and Human Induced Hazards (Vols. 1-2)

Editor: Chen Yong, *China Seismological Bureau, China*
eISBN: 978-1-84826-231-7, 978-1-84826-232-4
ISBN: 978-1-84826-681-0, 978-1-84826-682-7

Biodiversity Conservation and Habitat Management (Vols. 1-2)

Editors: Francesca Gherardi, Claudia Corti, and Manuela Gualtieri, *Universita di Firenze, Italy*
eISBN: 978-1-905839-20-9, 978-1-905839-21-6
ISBN: 978-1-84826-920-0, 978-1-84826-921-7

**Wildlife Conservation and Management in Africa

International Commission
President: Emmanuel K. Boon, *Free University of Brussels, Laarbeeklaan 103, Brussels, Belgium*
Vice Presidents:

Stephen Holness, *South African National Parks, South Africa*
Kai Schmidt-Soltau, *Cameroon*
Julius Kipng'etich, *Director of the Kenya Wildlife Service, Kenya*
Yaa Ntiamoa-Baidu, *Switzerland*
eISBN:
ISBN:

DEVELOPMENT AND ECONOMIC SCIENCES

Socioeconomic Development

Editor: Salustiano del Campo, *Universidad Complutense, Spain*
eISBN: 978-1-84826-085-6
ISBN: 978-1-84826-535-6

Welfare Economics & Sustainable Development (Vols. 1-2)

Editors: Yew-Kwang Ng and Ian Wills, *Monash University, Australia*
eISBN: 978-1-84826-009-2, 978-1-84826-010-8
ISBN: 978-1-84826-459-5, 978-1-84826-460-1

International Economics, Finance and Trade (Vols. 1-2)

Editor: Pasquale Michael Sgro, *Deakin University, Australia*
eISBN: 978-1-84826-184-6, 978-1-84826-185-3
ISBN: 978-1-84826-634-6, 978-1-84826-635-3

Global Transformations and World Futures (Vols. 1-2)

Editor: Sohail Tahir Inayatullah, *University of the Sunshine Coast, Maroochydore; Tamkang University, Taiwan; and, Queensland University of Technology, Australia*
eISBN: 978-1-84826-216-4, 978-1-84826-217-1
ISBN: 978-1-84826-666-7, 978-1-84826-667-4

Introduction to Sustainable Development

Editors: David V. J. Bell, *York University, Canada*
Yuk-kuen Annie Cheung, *York Centre for Applied Sustainability (YCAS); The University of Toronto, York University - Joint Centre for Asia Pacific Studies, Canada*
eISBN: 978-1-84826-222-5
ISBN: 978-1-84826-672-8

Principles of Sustainable Development (Vols. 1-3)

Editor: Giancarlo Barbiroli, *University of Bologna, Italy*
eISBN: 978-1-84826-079-5, 978-1-84826-080-1, 978-1-84826-081-8
ISBN: 978-1-84826-529-5, 978-1-84826-530-1, 978-1-84826-531-8

Dimensions of Sustainable Development (Vols. 1-2)

Editors: Kamaljit S. Bawa and Reinmar Seidler, *University of Massachusetts, Boston, USA*
eISBN: 978-1-84826-207-2, 978-1-84826-208-9
ISBN: 978-1-84826-657-5, 978-1-84826-658-2

Environment and Development (Vols. 1-2)

Editors: Teng Teng and Ding Yifan, *Chinese Academy of Social Sciences, China*
eISBN: 978-1-84826-270-6, 978-1-84826-271-3
ISBN: 978-1-84826-720-6, 978-1-84826-721-3

The Evolving Economics of War and Peace

Editors: James K. Galbraith, *Lyndon B. Johnson School of Public Affairs, The University of Texas at Austin, Jerome Levy Economics Institute and Chair of Economics Allied for Arms Reduction, USA*
Jurgen Brauer, *Augusta State University, USA*
Lucy L. Webster, *Economists Allied for Arms Reduction (ECAAR), USA*
eISBN: 978-1-84826-048-1
ISBN: 978-1-84826-498-4

Economics Interactions with Other Disciplines (Vols. 1-2)

Editor: John M. Gowdy, *Rensselaer Polytechnic Institute, USA*
eISBN: 978-1-84826-037-5, 978-1-84826-038-2
ISBN: 978-1-84826-487-8, 978-1-84826-488-5

International Development Law

Editor: A. F. Munir Maniruzzaman, *University of Kent at Canterbury, UK*
eISBN: 978-1-84826-314-7
ISBN:978-1-84826-764-0

INSTITUTIONAL AND INFRASTRUCTURAL RESOURCES

Human Settlement Development (Vols. 1-4)

Editor: Saskia Sassen, *The University of Chicago, USA*
eISBN: 978-1-84826-044-3, 978-1-84826-045-0, 978-1-84826-046-7, 978-1-84826-047-4
ISBN: 978-1-84826-494-6, 978-1-84826-495-3, 978-1-84826-496-0, 978-1-84826-497-7

Public Administration and Public Policy (Vols. 1-2)

Editor: Krishna K. Tummala, *Kansas State University, USA*
eISBN: 978-1-84826-064-1, 978-1-84826-065-8
ISBN: 978-1-84826-514-1, 978-1-84826-515-8

International Relations (Vols. 1-2)

Editors: Jarrod Wiener, *University of Kent at Canterbury, UK*
Robert A. Schrire,*University of Cape Town, South Africa*
eISBN: 978-1-84826-062-7, 978-1-84826-063-4
ISBN: 978-1-84826-512-7, 978-1-84826-513-4

International Law and Institutions

Editors: Aaron Schwabach, *Thomas Jefferson School of Law, USA*
Arthur John Cockfield, *Queen's University, Canada*
eISBN: 978-1-84826-078-8
ISBN: 978-1-84826-528-8

Institutional Issues Involving Ethics and Justice (Vols. 1-3)

Editor: Robert Charles Elliot, *Sunshine Coast University, Australia*
eISBN: 978-1-905839-14-8, 978-1-905839-15-5, 978-1-905839-16-2
ISBN: 978-1-84826-914-9, 978-1-84826-915-6, 978-1-84826-916-3

International Security, Peace, Development, and Environment (Vols. 1-2)

Editor: Ursula Oswald Spring, *National Autonomous University of Mexico-UNAM/CRIM,*
eISBN: 978-1-84826-082-5, 978-1-84826-083-2
ISBN: 978-1-84826-532-5, 978-1-84826-533-2

Conflict Resolution (Vols. 1-2)

Editor: Keith William Hipel, *University of Waterloo, Canada*
eISBN: 978-1-84826-120-4, 978-1-84826-121-1
ISBN: 978-1-84826-570-7, 978-1-84826-571-4

Democratic Global Governance

Editor: Irene Lyons Murphy, *Consultant, USA*
eISBN: 978-1-84826-181-5
ISBN: 978-1-84826-631-5

**National, Regional Institutions & Infrastructures

Editor: Neil Edward Harrison, *University of Wyoming, USA*
eISBN:
ISBN:

Conventions, Treaties and other Responses to Global Issues (Vols. 1-2)

Editor: Gabriela Maria Kutting, *University of Aberdeen, UK*

eISBN: 978-1-84826-233-1, 978-1-84826-234-8
ISBN: 978-1-84826-683-4, 978-1-84826-684-1

TECHNOLOGY, INFORMATION, AND SYSTEMS MANAGEMENT RESOURCES

Information Technology and Communications Resources for Sustainable Development

Editor: Ashok Jhunjhunwala, *Indian Institute of Technology, Madras, India*

eISBN: 978-1-84826-313-0
ISBN:978-1-84826-763-3

Systems Analysis and Modeling of Integrated World Systems (Vols. 1-2)

Editors: Veniamin N. Livchits, *Institute for Systems Analysis, Russian Academy of Sciences, Russia*
Vladislav V. Tokarev, *Moscow State University, Russia*

eISBN: 978-1-84826-088-7, 978-1-84826-089-4
ISBN: 978-1-84826-538-7, 978-1-84826-539-4

Systems Engineering and Management for Sustainable Development (Vols. 1-2)

Editor: Andrew P. Sage, *George Mason University, USA*

eISBN: 978-1-905839-00-1, 978-1-905839-01-8
ISBN: 978-1-84826-900-2, 978-1-84826-901-9

Knowledge Management, Organizational Intelligence and Learning, and Complexity (Vols. 1-3)

Editor: L. Douglas Kiel, *The University of Texas at Dallas, USA*

eISBN: 978-1-905839-11-7, 978-1-905839-12-4, 978-1-905839-13-1
ISBN: 978-1-84826-911-8, 978-1-84826-912-5, 978-1-84826-913-2

Science and Technology Policy (Vols. 1-2)

Editor: Rigas Arvanitis, *IRD - France & Zhongshan (Sun Yatsen) University, Guangzhou,*

eISBN: 978-1-84826-058-0, 978-1-84826-059-7
ISBN: 978-1-84826-508-0, 978-1-84826-509-7

Globalization of Technology

Editor: Prasada Reddy, *Lund University, Sweden*

eISBN: 978-1-84826-084-9
ISBN: 978-1-84826-534-9

Sustainable Built Environment (Vols. 1-2)

Editors: Fariborz Haghighat *Concordia University, Canada*
Jong-Jin Kim, *The University of Michigan, USA*

eISBN: 978-1-84826-060-3, 978-1-84826-061-0
ISBN: 978-1-84826-510-3, 978-1-84826-511-0

Integrated Global Models of Sustainable Development (Vols. 1-3)

Editor: Akira Onishi, *Centre for Global Modeling, Japan*

eISBN: 978-1-905839-17-9, 978-1-905839-18-6, 978-1-905839-19-3
ISBN: 978-1-84826-917-0, 978-1-84826-918-7, 978-1-84826-919-4

Artificial Intelligence

Editor: Joost Nico Kok, *Leiden University, The Netherlands.*

eISBN: 978-1-84826-125-9

ISBN: 978-1-84826-575-2

Computer Science and Engineering
Editors: Zainalabedin Navabi, *University of Tehran, Iran*
David R. Kaeli, *Northeastern University,*
eISBN: 978-1-84826-227-0
ISBN: 978-1-84826-677-3

System Dynamics (Vols. 1-2)
Editor: Yaman Barlas, *Bogazici University, Turkey*
eISBN: 978-1-84826-137-2, 978-1-84826-138-9
ISBN: 978-1-84826-587-5, 978-1-84826-588-2

LAND USE, LAND COVER AND SOIL SCIENCES

(Vols. 1-7)
Editor: Willy H. Verheye, *University of Gent, Belgium*
(Each Volume in the following covers a set of chapters, the total number of which is 78. The size of a chapter may vary from about 5000 words to about 30000 words.)
Volume 1: Land Cover, Land Use and the Global Change
eISBN: 978-1-84826-235-5
ISBN: 978-1-84826-685-8
Volume 2: Land Evaluation
eISBN: 978-1-84826-236-2
ISBN: 978-1-84826-686-5
Volume 3: Land Use Planning
eISBN: 978-1-84826-237-9
ISBN: 978-1-84826-687-2
Volume 4: Land Use Management and Case Studies
eISBN: 978-1-84826-238-6
ISBN: 978-1-84826-688-9
Volume 5: Dry Lands and Desertification
eISBN: 978-1-84826-239-3
ISBN: 978-1-84826-689-6
Volume 6: Soils and Soil Sciences - I
eISBN: 978-1-84826-240-9
ISBN: 978-1-84826-690-2
Volume 7: Soils and Soil Sciences - II
eISBN: 978-1-84826-241-6
ISBN: 978-1-84826-691-9

AREA STUDIES (REGIONAL SUSTAINABLE DEVELOPMENT REVIEWS)

Area Studies (Regional Sustainable Development Review): Africa (Vols. 1-2)
Editor: Emmanuel Kwesi Boon, *Free University Brussels, Belgium*
eISBN: 978-1-84826-066-5, 978-1-84826-067-2
ISBN: 978-1-84826-516-5, 978-1-84826-517-2

Area Studies (Regional Sustainable Development Review): Brazil
Editor: Luis Enrique Sanchez, *University of Sao Paulo, Brazil*
eISBN: 978-1-84826-171-6
ISBN: 978-1-84826-621-6

Area Studies (Regional Sustainable Development Review): USA (Vols. 1-2)
Editors: Lawrence C. Nkemdirim, *University of Calgary, Canada*
eISBN: 978-1-84826-069-6, 978-1-84826-070-2
ISBN: 978-1-84826-519-6, 978-1-84826-520-2

Area Studies (Regional Sustainable Development Review): China (Vols. 1-3)
Editor: Sun Honglie, *Chinese Academy of Sciences, China*
eISBN: 978-1-84826-071-9, 978-1-84826-072-6, 978-1-84826-073-3
ISBN: 978-1-84826-521-9, 978-1-84826-522-6, 978-1-84826-523-3

Area Studies (Regional Sustainable Development Review): Europe
Editors: Alexander Mather and John Bryden, *University of Aberdeen, UK*
eISBN: 978-1-84826-068-9
ISBN: 978-1-84826-518-9

Area Studies (Regional Sustainable Development Review): Japan
Editor: Yukio Himiyama, *Hokkaido University of Education, Japan*
eISBN: 978-1-84826-170-9
ISBN: 978-1-84826-620-9

Area Studies (Regional Sustainable Development Review): Russia (Vols. 1-2)
Editor: Nicolay Pavlovich Laverov, *RAS, Russia*
eISBN: 978-1-84826-074-0, 978-1-84826-075-7
ISBN: 978-1-84826-524-0, 978-1-84826-525-7

Caporaso, J.A.	USA	Dudley, W.	USA
Carabias, J.	Mexico	Edelstein, Norman M.	USA
Carpenter, A.	New Zealand	Eide, A.R.	USA
Carra, S.	Italy	El Ashry, M.	USA
Carroll, R.G.	USA	El Gohary, Mohamed I.	Egypt
Carta, S.	Italy	El Kassas, M.	Egypt
Carvalhal, Tania Franco	Brazil	El-Kholy, O.A.	Egypt
Catanzaro, E.	USA	Elliot, R.C.	Australia
Ceballos, G.	Mexico	El-Shaarawi, A. H.	Canada
Centi, Gabriele	Italy.	Eser, Semih	USA
Chaplin, R.	Canada	Etkin, N. L.	USA
Chapman, D.	Ireland	Farrell, R. V.	USA
Cheeseman, C.	UK	Favorsky, O.N.	Russia
Chen Jiaer	China	Feliu-Martinez, Juan M.	Spain
Chen, C.T.A.	Taiwan	Fellowes, M.	UK
Chen, Han Fu	China	Filar, J.A.	Australia
Cheng, Y.F.	Canada	Filho, Pedro de Alcantara Pessoa	Brazil
Cholakov, G. St.	Bulgaria		
Choppin, Gregory R.	USA	Filippov, G.	Russia
Choucri, N.	USA	Fisher, G.W.	USA
Christen, O.	Germany	Fleisher, D.H.	USA
Cilek, V.	Czech Republic	Fliedner, T.	Germany
Cockfield, A. J.	Canada	Ford, Paulette L.	USA
Coelho, Lísias	Brazil	Fortes, M.	Mexico
Colombo, U.	Italy	Fortes, M. D.	Philippines
Consoli, Fernando Louis	Brazil	Francis, J. F.	USA
Contrafatto, G.	South Africa	Frangopoulos, C.	Greece
Corazza, Giovanni E.	Italy	Frank , Roesch	Germany
Cordani, U.	Brazil	Freire, José Luiz de França	Brazil.
Corti, C.	Italy	Frieze, Paul A.	UK
Costello, Julio Alberto	Argentina	Frolov, K.V.	Russia
Costes, A.	France	Fuentes Gustavo A.	Mexico
Cote, R.P.	Canada	Fujii, T.	Japan
Cowling, R.M.	South Africa	Fukuoka, K.	Japan
Crothers, C.	New Zealand	Fukushima, Y.	Japan
Cui, Weicheng	China	Fuleky, G.	Hungary
Cumo, M.	Italy	Funabashi, M.	Japan
Czerny, Bozena	Poland	Galbraith, J.K.	USA
Da Silva, Andre Pires	USA	Gallart, F.	Spain
Dai Ruwei	China	Gallegos Crispulo	Spain
Daniels, J.L.	USA	Galles, C.D.	Argentina
Danilevich, Y.B.	Russia	Galles, Carlos Delfino	Argentina
Davies, A.M.	Israel	Gálvez, Julian Blanco	Spain
de Behar, Lisa Block	Uruguay	Gani, Rafiqul	Denmark
De Gaetano, Andrea	Italy	Ganti, Meenakshi	India
De Tombe, D.	The Netherlands	Gao Congjie	China
De Vivo, B.	Italy	Garzón, Francisco J. Morales	Colombia
Del Campo, S.	Spain	Gates, W. L.	USA
Del Claro, Kleber	Brazil	Geeraerts, G.	Belgium
Denemark, Robert A.	USA	Gellings, C.	USA
Dent, L.	Australia	Gerday, C.	Belgium
Derigs, U.	Germany	Gherardi, F.	Italy
Diamandopoulos A.	Greece.	Giegrich, J.	Germany
Dickinson, R.	USA	Glansdorff, N.	Belgium
Ding, Y.	China	Glazovsky, N.	Russia
Dixon, G.R.	UK	Gogus, Y.	Turkey
Djelal Kadir	USA	Goldberg, E.H.	USA
Djian, Jean-Michel	France	Goldemberg, J.	Brazil
Djian, J-M.	France	Goldstein, R.	USA
Doblare, Manuel	Spain	Golitsyn, G.S.	Russia
Doelle, Horst W.	Australia	Gollapudi, Maruthi Rao	India
Dolphin, D. W.	USA	Golley, F.	USA
Donarag, T. Y.	Malaysia	Gonzalez Osorio, L.,M.	Mexico
Donoso, M. C.	Panama	Gonzalez-Casanova, P.	Mexico
Dooge, J.C.I.	Ireland	Gorshkov, V.G.	Russia
Downing, T.G.	South Africa	Goulet, Denis	USA
Duarte, C.M.	Spain	Gowdy, J.M.	USA

Grabow, W.O.K.	South Africa	Jhunjhunwala A.	India
Grace, P.M.	Ireland	Jimeno, M.	Colombia
Gracia-Granda, Santiago	Spain	Jinsheng, G.	China
Grasemann, B.	Austria	Jintai, Ding	USA
Grasso, D.	USA	Johan, Elizabeth	Austria
Gray, Jeremy	UK	Jordaan, J.M.	South Africa
Green, D.G.	Australia	Jureckova, J.	Czech Republic
Gros, F.	France	Kaeli, D.R.	USA
Gruza, G.V.	Russia	Kakihana, H.	Japan
Gualtieri, M.	Italy	Kanwar, R. S.	USA
Guillermo, Ramirez	USA	Kanyar, Bela	Hungary
Guruswamy, L.D.	USA	Kao, D.T.	USA
Gyftopoulos, E.	USA	Kapitsa, Sergey	Russia
Haba, Hiromitsu	Japan	Karoly Suvegh	Hungary
Haddane, Brahim	Morocco	Kawabe, T.	Japan
Hafez, S.	Egypt	Kayis, A.B.	Australia
Haghighat, F.	Canada	Kazanjian, Arminee	Canada
Hallberg, R.O.	Sweden	Kerekes, J.	Hungary
Hamada, T.	USA	Kharkin, Valeri	Russia
Hammond, A.	USA	Khosla, A.	India
Hanninen, O.O.P.	Finland	Khublaryan, M. G.	Russia
Hansen, Vagn Lundsgaard	Denmark	Kiel, L. D.	USA
Hardege, J.D.	UK	Kim, J. J.	USA
Hardesty, D. L.	USA	Kim, T. J.	USA
Harkin, M. E.	USA	Kipng'etich, Julius	Kenya
Harl, N.H.	USA	Kirby, R.	South Africa
Harrison, N. E.	USA	Kitney, R.	UK
Hasan, S. E.	USA	Klir, G.	USA
Hashmi, M. U.	Pakistan	Klotz, S.	Germany
Hassan, P.	Pakistan	Knauth, Alfons	Germany
Hassan, S. Q.	USA	Kodama, Yoshiaki	Japan
Hassan, Y. A.	USA	Kok, J.N.	The Netherlands
Hasunuma, K.	Japan	Kolmer, L.	USA
Hatfield, J.L.	USA	Kondo, J.	Japan
Heden, C.-G.	Sweden	Kondratyev, K.Ya.	Russia
Helliwell, John R.	UK	Koteles Gyorgy J.	Hungary
Herrera, I.	Mexico	Kotlyakov, V.M.	Russia
Heurlin, B.	Denmark	Kralj, Slobodan	Croatia
Himiyama, Y.	Japan	Krashnoschokov, N.N.	Russia
Hipel, K.W.	Canada	Krawczyk, J.B.	New Zealand
Hirsch Hadorn, G.	Switzerland	Kubota, S.	Japan
Hoekstra, A. Y.	The Netherlands	Kumar, S.	Thailand
Holdgate, M.	UK	Kunnath, Sashi K.	USA
Holness, Stephen	South Africa	Kupitz, J.	Austria
Holton, J.R.	USA	Kurzhanski, A.B.	Russia
Holton, R.	Ireland	Kutting, G.M.	UK
Honglie, Sun	China	Kwon, Yong W.	USA
Horikawa, K.	Japan	Lagowski, J.J.	USA
Htun Nay	Myanmar	Lal, R.	USA
Hu Ying	China	Lam, Poh-Sang	USA
Huang, C.H.	Taiwan	Lara, F.	Mexico
Hudson, R.J.	Canada	Last, J.	Canada
Imura, H.	Japan	Lastovicka, J.	Czech Republic
Inayatullah, S. T.	Australia	Lasztity, R.	Hungary
Inyang, H. I.	USA	Lattanzio, John	Australia
Iribarne, O.	Argentina	Laverov, N.P.	Russia
Ishida, M.	Japan	Lee, Han Bee	Republic of Korea
Ishii, R.	Japan	Lee, S.	USA
Isla, F. I.	Argentina	Leemans, R.	Netherlands
Izrael, Y. A.	Russia	Lefevre, J. M.	France
Jablensky, Assen	Australia	Lefevre, T.	Thailand
Jackson, M.	UK	Leff, E.	Mexico
Jacobson, H. K.	USA	Leidl, R.	Germany
Jain, A.	India	Levashov, V. K.	Russia
Jakeman, A. J.	Australia	Levine, A.D.	USA
Jancso Gabor	Hungary	Li, Yuguo	China.
Jeyaratnam, J.	Singapore	Lichnerovicz, A.	France

Porembski, S.	*Germany*	Sinding-Larsen, R.	*Norway*
Prinn, R. G.	*USA*	Singeetham, Srinivasa Rao	*India*
Puel, J. P.	*France*	Singh, H.	*New Zealand*
Qi, L.	*China*	Singh, Prabhu N. P.	*UK*
Qian, Y.	*China*	Sirageldin, I.	*USA*
Quesada, Mauricio	*Mexico.*	Siva Sivapalan, M.	*Australia*
Qureshy, M. N.	*USA*	Sivov, N.A.	*Russia*
Rajeswar, J.	*India*	Sixto Malato, Rodríguez	*Spain*
Ramallo, L.I.	*Spain*	Skogseid, H.	*Norway*
Ramirez, Alonso	*USA*	Smeers, Y.	*Belgium*
Ramirez, Nelson	*Venezuela*	Smets, B.	*USA*
Rao, C. N. R.	*India*	Smith, S.	*UK*
Rao, C.R.	*USA*	Socanac, Lelija	*Croatia*
Rao, Ganti Prasada	*India*	Sollars, C.J.	*UK*
Rapport, D.J.	*Canada.*	Sonntag, H.	*Venezuela*
Rasool, I.	*France*	Soria Escoms, B.	*Spain*
Raumolin, J.	*Finland*	Sorour, K.	*Egypt*
Rawlings, R.D.	*UK*	Southwood, R.	*UK*
Reddy, P.	*Sweden*	Speight James G.	*USA*
Reza, F.	*Canada*	Squires, V.R.	*Australia*
Rheinberger, Hans-Jörg	*Germany*	Squires, Victor R.	*Australia*
Ricco, Bruno	*Italy*	Stabell, Rolf	*Norway*
Rico-Gray, Victor	*México*	Stepp, John Richard	*USA*
Rios, Mônica Palácios	*México,*	Stuewe, K.	*Austria*
Rodrigues, Silvio Carlos	*Brazil*	Sujoldzic, A.	*Croatia*
Rodriguez, L. F.	*USA*	Sulayem, M.	*Saudi Arabia*
Rokem, J. Stefan	*Israel.*	Sullivan, L.	*Australia*
Rúbio, Scarano Fábio	*Brazil*	Summy, Ralph V.	*Australia*
Rudan, P.	*Croatia*	Susheela, A.K.	*India*
Ruigómez, José Ma Goicolea	*Spain*	Sydow, A.	*Germany*
Rydzynski, K. J.	*Poland*	Szczerban, J.	*Poland*
Sabljic, A.	*Croatia*	Takahasi, Y.	*Japan*
Saccomandi, Giuseppe	*Italy*	Takeda, G.	*Japan*
Safran, P.	*Philippines*	Takle, G.S.	*USA*
Sajeev,, V.	*India .*	Tao, W-Q	*China*
Sala, M.	*Spain*	Tarasova, N. P.	*Russia*
Sala, O.	*Argentina*	Tarlock, A.D.	*USA*
Salim, E.	*Indonesia*	Teller, E.	*USA*
Sampaio, Marcus Vinicius	*Brazil*	Teng, T.	*China*
Sanchez, L.E.	*Brazil*	Thayamballi, Anil K.	*USA*
Santamouris, M.	*Greece*	Thiessen, Kathleen M.	*USA*
Santiago Gracia Granda	*Spain*	Ting, K.C.	*USA*
Sassen, S.	*USA*	Titli, A.	*France*
Satoh, T.	*Japan*	Toschenko, J.T.	*Russia*
Savenije, H.H.G.	*The Netherlands*	Tsipenyuk, Y.M.	*Russia*
Sayers, B. McA.	*UK*	Türmen, Tomris	*Turkey*
Scarano, Fábio Rúbio	*Brazil*	Ul Haq, M	*Pakistan*
Schmidt-Soltau, Kai	*Cameroon*	Unbehauen, H.D.	*Germany*
Schrire, R.A.	*South Africa*	Valero-Capilla, A.	*Spain*
Schwabach, A.	*USA*	van den Haute, Peter	*Belgium*
Seidler, R.	*USA*	van der Beken, A.	*Belgium*
Sekiguchi, M.	*Japan*	van Wyk, J.A.	*South Africa*
Sen, G.	*India*	Vanek, Tomas	*Czech Republic.*
Serageldine, I.	*USA*	Varga, Kálmán	*Hungary*
Serneri, Simone Neri	*Italy*	Vazquez, C.	*Mexico*
Sgro, P.M.	*Australia*	Verhasselt, Y. L. G.	*Belgium*
Shaib, B.	*Nigeria*	Verheye, Willy	*Belgium*
Shaidurov, V.V.	*Russia*	Viertl, R.	*Austria*
Shen, Hayley	*USA*	Vigneswaran, S.	*Australia*
Shidong, Z.	*China*	Villarnovo, Glória Carrion	*México*
Shiklomanov, I.A.	*Russia*	Vishnevsky, A.G.	*Russia*
Shpilrain, E.E.	*Russia*	Volgy, T.J.	*USA*
Shuttleworth, W.J.	*USA*	Von Weizsacker, E.U.	*Germany*
Siebel, Wolfgang	*Germany*	Wallenius, Maria	*Germany*
Silva, Márcio Seligmann	*Brazil*	Wang, Ge (George)	*USA*
Silveira, L.	*Uruguay*	Wang, Mingxing	*China*
Silvio, Carlos Rodrigues	*Brazil*	Wang, W.C.	*USA*

Wang, Xingyu	China	Wohnlich, S.	Germany
Wang, Yuqing	USA	Wojnarovits, Laszlo	Hungary
Webster, L.	USA	Wojtczak, A.M.	USA
Weder, R.	Mexico	Wolanski, E.	Australia
Weffort, F.	Brazil	Woldai, A.	Eritrea
Wei, Zhou	Hungary	Wong, K.P.	Australia
Weiss, T.G.	USA	Xila S. Liu	China
Welch, R.M.	USA	Yaa, Ntiamoa-Baidu	Switzerland
Wiener, J.	UK	Yagodin, G.A.	Russia
Williams, Paul D.	UK	Yam, Kit Keith L.	USA
Wills, I.	Australia	Yan, S.	China
Wilson, P.A.	UK	Yi, Zeng	USA
Wilson, S.R.	Australia	Zhang, Wei-Bin	Japan

UNESCO-EOLSS JOINT COMMITTEE

A UNESCO-EOLSS Joint Committee was established with the objectives of (a) seeking, selecting, inviting and appointing Honorary Theme Editors (HTEs)/International Commissions for the Themes, (b) providing assistance to them, (c) obtaining appropriate contributions for the different levels of the Encyclopedia, and (d) monitoring the text development. The membership of the Committee is as follows:

Badran, A.	*President, Arab Academy of Sciences, (Co-Chairman)*
Al Gobaisi, D.	*EOLSS Editor-in-Chief (Co-Chairman)*
El Tayeb, M.	*Former Director, Division for Science Policy & Sustainable Development, UNESCO, (Secretary)*
Tolba, M.K.	*President, International Center for Environment and Development, Former Executive Director of UNEP*
Sage, A.P.	*George Mason University, USA (Chairman, EOLSS Configuration Control Board)*
Marchuk, G.I.	*Russian Academy of Sciences*
Szollosi-Nagy, A.	*Rector, UNESCO-IHE Institute for Water Education*
Chesters, G. (Deceased)	*University of Wisconsin, USA*
Johns, A.T.	*University of Bath, UK*
Lundberg, H.D.	*Sweden*
Younes, T.	*International Union of Biological Sciences, France*
Dempsey, J.	*UK*
Ganti, Prasada Rao	*India*
Joundi S.	*France*
Makkawi, B.	*Sudan*
Woldai, A.	*Eritrea*
Agoshkov, V.I.	*Russia*
Hornby, R. J.	*UK*
Wall, G.	*Sweden*
Watt, H.M.	*USA*
Kotchetkov, V.	*Consultant, UNESCO*
Al-Radif, A.	*Canada*
Sasson, A.	*Morocco*
Bruk, S.	*Consultant UNESCO*

Secretariat

Huynh, H.	*Assistant Project Coordinator, SC/PSD, UNESCO*
Barbash A.	*Systems Manager, SC/PSD, UNESCO*

Further details may be obtained from

www.eolss.net

helpdesk@eolssonline.net

DISCLAIMER